Grounded Theory

A Reader
for Researchers, Students, Faculty and Others

by

Dan Remenyi PhD

Grounded Theory
A Reader
For Researcher, Students, Faculty and Others
First Edition, September 2013
Second Edition, July 2014

ISBN: 978-1-909507-90-6

Published by: Academic Conferences and Publishing International Limited, Reading, RG4 9SJ, United Kingdom,
info@academic-publishing.org

Printed by Lightning Source

Available from www.academic-bookshop.com

About the author

Dr Dan Remenyi formerly a Visiting Professor specialising in research methodology at seven universities in four countries over the past 20 years. He continues to write, teach and research in both research methodology and the sociology of research. He conducts seminars on topics related to improving effective academic research and obtaining better research results. One of his areas of specialism is qualitative research and how it may be enhanced using a Grounded Theory approach. He is on the editorial board of a number of academic journals. He is also on the executive committees of several European and International conferences. His research has been published in some 50 peer reviewed papers and he has had some 30 text books published. Some of his books have been translated into Chinese, Japanese and Romanian. He holds a B Soc Sc, MBA and PhD.

Contents

Preface

From the first time I encountered Grounded Theory some twenty years ago I knew it was a most important issue for researchers in social science. It was clear to me at that time because I was looking to find some sort of formula to apply to my doctoral research, and none of the people with whom I had contact were able to express, in terms understandable to a novice researcher, the essence of academic research. I felt that I was floundering. There were lots of people with doctorates around who had wise advice, but I could not at that time relate to what these individuals were saying.

Grounded Theory was a lifeline for me in two important respects. Firstly it offered me some tools and processes and for these I was grateful. I could use these to move forward through the processes of academic research. But more importantly it made me, a rather lowly doctoral degree candidate, feel that I understood something about the nature of the activities which were necessary for competent research and that I could then make a contribution to my field of study.

As is mentioned in both the first and second chapters of this reader, Grounded Theory can be really quite fussy and the debates between the leading proponents can be distracting. But this does not in any way detract from the real strength of Grounded Theory, which is that it takes the lid off the "black box" of qualitative research and allows anyone who wants to look inside to see what goes on. It is for that reason that I believe all social science researchers should know about this subject. Being concerned as to whether a particular book or seminar addresses the Glaser and Strauss or the Corbin and Strauss approach also seems to me to miss the point. Every implementation of Grounded Theory will be different, and so it should.

However, this is not to say that some of the disputes which have arisen from the Glaser and Strauss or the Corbin and Strauss debates have not been instructive. Some of them have been valuable. Issues related to the so-called forcing of theory and the role of literature are enlightening.

To be a successful researcher it is necessary to understand theory, and it is sad to say that many novice researchers do not. Thus there is a chapter (previously published as a paper) on theory and theoretical research. Equally distressing is the fact that, knowledge of what data really is can be absent from many books, courses and seminars on research and this is addressed. Finally the philosophical stance of much of social science and especially research based on Grounded Theory is pragmatism and this also is poorly understood.

This Reader offers insights into these topics. There is no suggestion made that these topics are covered in any definitive way. It is not the function of a Reader to provide depth but to equip those who consult this book with a comprehensive starting point to move forward to a deeper level of knowledge.

Dan Remenyi PhD
dan.remenyi@gmail.com

How to use this book

One of the characteristics of this Reader is that it is a dense book containing highly focused chapters, some of which are based on papers written for detailed study.

To understand the nature of Grounded Theory the first chapter needs to be read especially carefully. It is a summary of the main arguments related to Grounded Theory as well as a high level guide to the work involved with this method.

The second chapter puts the whole idea of methodology and methods into a context which recognises their importance but which at the same time points out that there can be over emphasis on research methodology. This chapter is a direct follow-on from chapter one.

The subsequent three chapters are discussions on Theory, Data and Pragmatism which may be read independently. The next two chapters address the functioning of Grounded Theory while the final chapter asks a searching philosophical question.

A substantial glossary is provided for reference.

This Reader is designed to help a researcher get going in Grounded Theory or to understand at a deeper level some of the issues involved in qualitative research. But understanding the issues described herein are only a start on an exciting journey for academic researchers. The Reader is in no way a definitive account of this subject. It is written to be a useful launch into the intriguing field of Grounded Theory.

Acknowledgements

This book, like so many in the field of research methods, is a work in progress.

The thinking behind this book started many years ago when I first became aware of Grounded Theory, when the long journey to understand what it was about began.

Grounded theory is not often taught well. For that matter many aspects of research methods are left for students to find their own understanding by trial and error.

I have been helped in my academic research journey by too many people to name anyone specifically and therefore I need to say that without the intellectual engagement of my colleagues and students in this subject this particular Reader could not have been produced.

I am grateful for all the help and encouragement I have received over the years.

Prelude - Thinking about the central dilemma in teaching academic research

It is increasingly accepted that having students find out for themselves is pedagogically more effective than telling them something, and this insight has been reflected in the more extensive use of research as a central part of the process of education in many fields of study.

When research is conducted well it is a rewarding activity and it brings two levels of satisfaction. The first is that the researcher is pleased to have the answer to the question and the second is that the researcher can feel satisfied that he or she has mastered the art or science of research (at least to a certain extent).

However in this there is a dilemma, which can be expressed as follows: Should academics allow their students to learn how to research by themselves? It has just been said that finding out for oneself is the best way to learn. Although this is true it has two disadvantages. Firstly, it often does take more time to explore new ideas, concepts and processes on one's own and secondly, research is a subject area in which it is possible to take a wrong turn that is not obvious and thus find out later that much work has been done which will not bear any fruit. Therefore is it appropriate to advise novice researchers how to go about their research?

How can the question, How should the research be conducted? be answered? Most academics have traditionally shied away from this question. In the past novice researchers have often been left largely to their own devices to learn by trial and error, and although this does in fact work some novice researchers are discouraged and do not complete the projects they originally started.

So, how should academic research be conducted? At a high or conceptual level the answer to this question is, find a suitable problem or research question, understand what data is required to be able to obtain a comprehension of the issues involved, produce a data acquisition plan, access and manage the data, process it appropriately, seek more data if required, and then interpret the results. Unfortunately an answer at this high level is often of little help to a novice researcher. Much more detailed advice and guidance is normally required to get started on the right footing and that is where Grounded Theory can play an important role. To a large extent Grounded Theory can provides step by step advice on how each part of the research should be conducted and those who are new to academic research find this reassuring.

But the detail is not of any real value unless the philosophical underpinning of the research is adequately understood. Grounded Theory stands four-square behind Pragmatism. Pragmatism, which is discussed later, is a multifaceted concept but suffice it to say here that pragmatism asserts that knowledge should be thought of as a work in progress. How similar this is to Giddens (1990) who said, "In science, nothing is certain, and nothing can be proven, even if scientific endeavour provides us with the most dependable information about the world to which we can aspire"?

But to benefit from Grounded Theory you don't have to be a devoted follower and conscientious observer of all the different facets of the method. Researchers are free to use Grounded Theory ideas to help them understand the processes of research and to pick and choose which aspects of the method they wish to employ in their own research.

Do remember that every academic research project is different in how it is executed. All academic research projects are the same in their objective in that they need to add something of value to the body of knowledge and the researcher needs to do this convincingly, but the way this is achieved will be uniquely personal.

What has to be consistent across all academic research projects is clear evidence of scholarship and a significant contribution to both theory and to the improvement of practice.

1

What is Grounded Theory and what does it do for Social Science?

How important is Grounded Theory to social science research?

1.1 Introduction

This chapter provides some of the answers to the question *What is Grounded Theory and what does it do for social science?* This is an important question and in the following pages a high level overview is presented, the objectives of which are to introduce the language and the rhetoric of the method. Even after many years of study and practice researchers sometimes report that they have just come to realise the importance and the effect of some dimension of Grounded Theory.

Grounded Theory not only offers a method by which social science research may be rigorously conducted but it also provides a more general explanation and understanding of how qualitative research operates. The primary purpose of the original book written by Glaser and Strauss was not to provide a step-by-step guide to theory generation but rather to establish the principle that by using an appropriate method researchers could make a valuable theoretical contribution to their field of study. In so doing Glaser and Strauss took the lid off what had been the "black box" of academic research in the Social Sciences and allowed some of the inner workings of this subject to be viewed and assessed. This was a major contribution to enhancing the general understanding among academics of social science research.

1.2   The Success of Grounded Theory

Grounded Theory is the most successful innovation in academic research in the social sciences of the 20th century, being both original and enduring. It liberated or democratised research in a way that no other innovation had done before and thus it allowed even newcomers to research an opportunity to argue in a structured way that they had made a theoretical contribution in their field of study. Grounded Theory made the nature of qualitative academic research visible and comprehensible to those who wish to look closely at its processes. In so doing it enhanced the credibility of this research approach and strengthened its claim to be authentic and scientific.

Before Grounded Theory, the scope of the research activities for doctoral degree candidates and other novices to academic research was considered to be that of theory testers (also sometimes called puzzle solvers). Therefore the best that a novice researcher could hope to achieve when testing established theory was to reject an hypothesis and thereby be able to make some relatively minor modification to a theory. Prior to Grounded Theory it was generally held that theory creation was reserved for the so-called intellectual capitalist, a term which is now somewhat out of fashion.   Another term used to describe theory from former times was Grand Theory. Grand Theory would have been postulated by authorities like Smith, Durkheim,

Marx, Weber and other academic masters. Mere mortals stood in the shadow of these giants.

1.3 Theory Generation and not Theory Testing

Glaser and Strauss demonstrated relatively convincingly that if a rigorous method was carefully followed there was no reason why novice researchers could not make a theoretical contribution and perhaps even an important theoretical contribution to their field of study. It needs to be noted that this Grounded Theory is not designed for theory testing but for theory creation. These are normally perceived as two different research activities. Anyone who suggests that Grounded Theory could be used for hypothesis testing is mistaken[1] Thus, when using Grounded Theory the research question would be in the following form: *What theory would offer a reasonable understanding of the phenomenon of interest?* The research question is always central in academic research and when Grounded Theory is employed the research question will be answered by proposing a theory or a theoretical conjecture.

The nature of theory is not well understood, even by some academics, and this topic requires specific and detailed attention by the Grounded Theory researcher. In the context of the research described in this chapter a useful definition of theory is that:

---

[1] There are aspects of the Grounded Theory method which are of considerable value to theory i.e. hypothesis testing in an academic environment but this is not the primary purpose for which this approach was developed.

A theory is systematically organised knowledge applicable in a relatively wide variety of circumstances, using a system of assumptions, accepted principles and rules of procedure devised to analyse, predict or otherwise explain the nature or behaviour of a specified set of phenomena. But it is also often simply the best explanation which is available at that time.

This subject is more fully addressed in Chapter 3.

Although most Grounded Theory enthusiasts emphasise the theory creation nature of the method, many of the issues raised by Glaser and Strauss in their 1967 book, *The Discovery of Grounded Theory* provide good practice for academic researchers, especially qualitative researchers, irrespective of the objectives of their research.

1.4   Grounded Theory and Induction

Grounded Theory is also important because it recognises the contribution which is made to scientific discovery by the process of induction. Glaser and Strauss (1967) define Grounded Theory as an induction based method as follows:-

Grounded theory is an inductive, theory discovery methodology that allows the researcher to develop a theoretical account of the general features of a topic while simultaneously grounding the account in empirical observations or data.

By induction is meant the process of abstracting from an incident or phenomenon, or a series of incidents or phenomena, a general principle or theory which will describe the nature of the incident or phenomenon incorporating the issues, variables or constructs present in the incident. Induction can also be described as the intellectual process or a method of inference in which the researcher moves from data to theory[2].

Induction is not an un-problematic concept. It has detractors and important detractors at that. David Hume, one of the distinguished thinkers of the Enlightenment, questioned the validity of any science based on induction. But in recent times perhaps the most distinguished detractor of induction was Karl Popper. Popper pointed out that the logic which is normally associated with induction is flawed. His argument was that with empirical research, no matter how much data is collected in support of a proposed theory, it is possible that the next data item acquired could contradict that theory[3]. This insight led Popper to postulate the notion of falsification for which he became famous. Although data cannot be used to "prove" a theory it can always be used to falsify a theory. This leads directly to acknowledging the conjectural nature of knowledge and requires us to appreciate that our knowledge at any point in time should be considered to be our best understanding at that point in time. It is not possible to predict how our understanding will evolve. What can be said is that our cognitive capacity has been evolving for millennia and

---

[2] Induction is often contrasted with deduction where the researcher begins with a theory and moves from that position to creating hypotheses and then from there to collecting data with which to attempt to reject the hypotheses. Deduction has been described as moving from theory to data.
[3] Although in principle this is a logical objection to induction, in practice there are always outliers and even anomalies.

there is no a priori reason to suppose that it will not continue to so do.

Einstein similarly argued that there was no scientific method or logic in induction. There is no scientific basis for believing that just because something has occurred in the past, it will occur again in the future. There is no doubting the logic of these statements. But nonetheless induction is a particularly important inference method for scientists who have been able to use it to produce many effective theories over a large number of centuries. Scientists generally do not have difficulty in 'living with' the inherent uncertainty associated with any finding based on induction. Some researchers are surprised by how little respect is given to induction by philosophers, as an approach to science. This is, of course, a paradox as at least superficially it appears to contradict the fact that induction is so commonly employed. There is what is sometimes referred to as the psychological dimension to induction. Human life constantly depends upon a series of assumptions relating to the belief that tomorrow will largely be the same as today, at least as far as the laws of physical and life sciences are concerned. These assumptions relate to matters such as the sun will rise and set at similar times to when it did today; there will be two high and low tides each day; when the cooker is powered up the plates will become hot and it will work; if hot smoking oil is splashed on someone's hand it will hurt; when water is poured into a glass it will not leak (unless there is a crack in the glass); when a car drives on a road it will not open up and swallow the vehicle. These are all induction based assumptions without which life would be impossible. What is important to note is that there is no intrinsic logical reason why any of these assumptions should be correct. Nonetheless these assumptions are very often right and make life possible as we know it. Thus, some scientists argue that there is a psychological reason why the human mind turns to induction because although it may not be based on logic there is experience which leads to belief in its efficaciousness.

Induction is an acceptable approach to academic research provided that the researcher is aware of and makes clear the contingent nature of the knowledge produced. Without this caveat induction does not make logical sense.

1.5 Abduction

Grounded Theory also directs our attention again to the issue of abduction, which when used with Grounded Theory is an approach to inference. Abduction is an ugly word which was coined by the man who is credited with the establishment of Pragmatism, Charles Sanders Pierce, to describe another method of research inference. It is not a simple matter to define or to describe abduction. A researcher employs abduction when he or she proposes a connection or a pattern between different sets of data or evidence without being aware of all that could be learnt about the context and the nature of the data sets. If understood in this way abduction may be seen as a special case of induction. In common parlance one can describe abduction as a type of informed guessing. It may also be described as that which occurs when a researcher serendipitously has a flash of insight. There is no guarantee that guesses or flashes of insight would lead to a correct connection being made and they should be seen only as a starting point which would cause the researcher to make further investigations. Although researchers are often reluctant to describe their work in terms of guesses and flashes of insight it is commonly known that these "methods" have played and continue to play a significant role, in the form of a suggested line of enquiry, in much scientific research. Thus even if an apple did fall on Newton's head and this led to his wondering how the apple's fall and the moon's orbit could be connected, there was still much scientific work to be done before Newton's Laws of Motion could be postulated.

The following is an example of abductive thinking:-

Here are the facts. In June 2013 the BBC reported that there had been a bizarre incident in the Algarve in Portugal concerning British Tourists gambling. Gambling is illegal in that country. A British Pub had been running a Bingo Game. The players had been paying one Euro each per game and the prizes which the Pub offered to the winners were biscuits. The police raided this den of iniquity and stopped the game. The gamblers were arrested and charged and then appeared in court. The judge pronounced sentences which included three months in prison, suspended and fines of Euros 300. Those other drinkers in the pub who were not gambling but only observing the illegal proceedings were fined Euros 150.

The landlady of the offending establishment was also sentenced to three months in jail suspended and fined Euros 700.

The news headline here was that the Portuguese Police heavy handed reaction to law breakers. British public opin-

ion would of course respect Portuguese Law but would be more inclined to see this incident as requiring the people involved to receive cautioning rather than criminal sentences.

Other possibly related events:- *The Portuguese justice system is generally poorly regarded in Great Britain after the Madeleine McCann affair. Madeleine McCann was a 4 year old when she was taken from her parent's apartment in a holiday resort in Portugal. She has never been found. It is generally said that at best the Portuguese police could be regarded as ineffective. There has been more criticism of the Portuguese police handling of this case recently when the Metropolitan Police in London reopened the Madeleine McCann case.*

Here is a possible explanation. It seems that this crime stopping intervention on the Algarve might be the Portuguese Police showing the British that they are a serious power in their land. Whatever the British might think of the Portuguese Police, Portuguese Law cannot be flaunted in Portugal. Of course if this is indeed the case then it works directly against their own interests and makes the Portuguese Police and the Portuguese justice system look even more inept than they previously did.

This explanation, based on an abductive inference, is pure speculation and it is now necessary to explore in detail the circumstances around the incident in the English pub. It would also be important to examine whether the recent reopening of the McCann affair really caused offence in Portugal and whether this could have had a direct impact on police action in the Algarve.

1.6 Dataism and data

The term Dataism is sometimes used to describe the type of approach which is used in Grounded Theory research. This term is used to signal the paramount importance which is attributed to data by Grounded Theory researchers. Of course, data is always important in academic research but in the process of theory development through the Grounded Theory Method data takes on an even greater role than normal. The researcher is expected to collect a material amount of data. Although the data collected has to be carefully focused on the research question, the suggestion is that unstructured interviews may lead to a richer harvest of data[4]. It is not formally required that the researcher engages in a prolonged immersion with the informants as is common in ethnographic studies, but the researcher needs to have a solid understanding of the phenomenon or issues being studied. Any research data recorded has to be subjected to some process of continuous comparison with other data collected. To facilitate this constant comparison the researcher needs to work on his or her data as soon as possible after acquisition and use the knowledge developed to improve the next round of data acquisition.

The concept of continuous comparison is made much of by Grounded Theory researchers and it is sometimes incorporated in the definition

---

[4] As a general rule university Ethics committees are becoming less prepared for Ethics protocols to include unstructured interviews. They increasingly want a list of questions which will be asked and they expect researchers to adhere to them.

of this approach to research. However, on reflection academic research of all sorts will employ to some extent a continuous comparison frame of mind towards data collection. The data collected during Grounded Theory research has to be sifted through carefully and the researcher is directly responsible for the integrity of this data. In addition the researcher needs to be constantly aware of where the data that is being collected is taking the research argument. This does not mean that the researcher can or should only seek out data which will support his or her argument. In fact it can be the opposite and the researcher will sometimes try to find evidence which may contradict the apparent findings. One advantage of this continuous awareness of the data as it is being collected is that it allows the researcher to identify outliers in real time. Identifying and understanding outliers is essential for competent research.

When the theory is being formulated the researcher needs to keep a tight focus on how the theoretical conjectures being created are reflected in the data. If any part of the theory is questioned then the researcher needs to be able to point to that aspect of the data which underpins his or her theoretical interpretation. This is not unique to Grounded Theory, but Grounded Theorists often point to it as a fundamental precept in the Grounded Theory Method.

Chapter 3 addresses some of the issues related to the nature of data and how it may be perceived by researchers. In the case of Grounded Theory, Glaser and Strauss famously remain rather vague on this issue. This is not surprising as many researchers seem to think that the concept of data does not require much, if any, thought. "It is so obvious, isn't it, what data is", remark many researchers. But this is not the case. Alvesson and Skoldberg (2009) point out:

> *In no small number of cases, the word 'data' seems to mean whatever Glaser and Strauss arbitrarily chose it to mean. Generally speaking, then, data in grounded theory can be described in vague terms as something empirical, often some event, often in the form of an incident, often in the form of some social interaction.*

What is important to appreciate is that this is only one type or dimension of data which was the most appropriate for the uses to which Grounded Theory was put by the masters, i.e. Glaser and Strauss. Glaser famously pointed out that *all is data* and although this is an unfortunately glib expression, it does point to the fact that researchers always have to be on the lookout for any data which can contribute to the objectives of the research.

An important aspect of dataism is that the processes of data collection and data understanding (and thus to some extent interpretation) are to a large extent integrated. This allows the researcher to engage in constant improvement of his or her technique as the data collection proceeds. The data collection activity becomes a learning process in its own right. This represents a considerable advantage over the collection of data through a measuring instrument such as a questionnaire. If there is any problem in the questionnaire the researcher has to live with it for the rest of the research. Repeating the data collection process with an improved questionnaire is often not an option.

1.7   A Research Protocol

A research protocol is always helpful to a researcher, if for no other reason than it requires careful thought to be given to the research processes necessary to achieve the project objective. In fact many universities now insist that a research protocol should be part of the ethics application and this normally needs to be approved before the research can commence.

Ethics committees are generally uncomfortable in allowing much flexibility when approving a research project. The approval process is designed so that the ethics committee will be fully briefed as to what is required for the research and that they give their permission for only those activities. Thus there can be a conflict as Grounded Theory emphasises flexibility and this means that the researcher needs to produce a research protocol which can allow new sources of data to be explored if the opportunity arises. This could mean that the re-searcher may have to seek further ethics approval as the research project proceeds and from the researchers point of view this is an unwelcome complication.

1.8   Theoretical Sampling

Grounded Theorists will employ Theoretical Sampling techniques. The concept of theoretical sampling often presents challenges to re-searchers, especially those who have been schooled in the type of sampling techniques discussed by statisticians. In rather simple terms theoretical sampling may be seen as a type of purposeful sampling where the researchers are seeking incidents of the phenomenon they are studying and which will supply data that they regard to be useful to further describe the phenomena of interest. The purpose of this is to enhance the understanding of the categories and concepts that are developing. Theoretical sampling also requires the researcher to be alert to acquiring more data in support of the theory which is be-

ing developed in his or her mind. If data is found which does not support the theory, especially contradictory evidence, this is also considered useful and should prompt the researcher to explore further as
to why there are contradictions. Researchers always have to be cog-
nizant that there will inevitably be outliers and anomalies.

When acquiring data no preference should be established for one or
more data source. Although much Grounded Theory relies on inter-
viewing, researchers need to be sensitive to as many different
sources of data as possible. There are many different ways of acquiring data and there are often valuable caches of data which can be
pursued and used.

Theoretical sampling is quite different to statistical sampling which is
used to collect data for the purposes of measuring variables or constructs in a population.

Researchers who are new to Grounded Theory sometimes need reas-
surance that random sampling is not required. Grounded Theory researchers are often looking for good practice and are therefore not
interested in finding the type of evidence which they might encounter if their informants were to be chosen by any random sampling
technique. It is for this reason that theoretical sampling has been described as having some of the characteristics of triangulation.

There is some debate about the use of the word sampling in this context. The word sample normally refers to a researcher taking a small
number of items from a large population. Traditionally the researcher
will have some idea of how big the whole population is and he or she
will be looking for items which can be said to represent in some

sense the whole population. But in theoretical sampling the re-searcher is not looking towards finding situations which represent a whole population. The researcher is seeking incidents of a particular phenomenon that will throw light on a theory which can be used to explain the phenomenon. It is because of this that grounded theo-rists distinguish between substantive theory and formal theory. In general, substantive theory will often have a low level of abstraction and formal theory is the name we use when a higher level of abstrac-tion is involved[5]. These concepts will be further explored later.

1.9   Data or Theoretical Saturation

Grounded Theory requires a substantial amount of data and it is quite difficult to provide guidelines as to how much data this actually represents. This is not possible to determine a priori. The research question and the context are the primary drivers which determine how much data will be required and thus the concept of Data or Theoretical Saturation is used.

The concept of Data or Theoretical Saturation is straightforward as it is defined as being reached when the researcher no longer finds new facts or figures or ideas being provided by additional data sources. This is of course rather subjective and it is often challenged by re-viewers and examiners.

Sometimes a researcher will believe that he or she has reached satu-ration but will subsequently, during analysis or theory emergence,

---

[5] It is also correct to say that formal theory will generally have a greater scope and possibly more generalisation then substantive theory.

realise that more data would be helpful. Under such circumstances the researcher should return to data acquisition. If this occurs it may be the result of inadequate research design and this may mean changes to the research design and also a new Ethics Protocol application.

1.10 Data and Reflection thereon

Some aphorisms are untrue and one of the most obvious of these is the saying, "the facts speak for themselves". The facts or data (a term preferred by academic researchers) are always subjected to interpretation and hopefully an interpretation is arrived at after some careful thought or reflection. Data without being evaluated through appropriate reflection is unlikely to be able to answer the research question. In the context of data within the Grounded Theory research paradigm, reflection requires the researcher to think carefully about issues related to: what data was sought, how was it acquired, was the data comprehensive, has one data set pointed to the need for another and where are these research activities leading? It is interesting to recall how reflection was described specifically as a tool by Sherlock Holmes. Conan Doyle famously had Holmes talk about certain problems being a "three pipe problem". Perhaps that implied a whole evening in deep thought.

Unfortunately, sometimes reflection can end up with the researcher justifying his or her approach and when this happens there is little value to be obtained. Reflection does not occur naturally and researchers need to have a structure in which to reflect. There are different approaches to reflection and some checklists have been developed. However, the essence of these is that the researcher continually needs to question the efficaciousness or utility of the research activities.

Reflection permeates all aspects of research and so it will still be required with regard to making sense of the data. Here more structure is often required and in creating this structure the researcher needs to decide how to group, sort or arrange the data in order to come to an understanding of and an answer to the research question. This usually means that the data has to be consolidated in some form in order to facilitate this process. This is at the heart of much qualitative data analysis and is not a simple matter. The first step in this is sometimes data coding which is the process of labelling concepts arising in the data, but before data coding can begin the researcher has to identify the concepts which should be coded[6].

There is no mechanistic formula which could be offered to assist a researcher in this respect and it is undoubtedly the case that two different researchers will identify different issues to code. The same researcher may at different times wish to use different codes for the same type of concept mentioned in the text. Thus, there are always issues of reliability and consistency and coding needs to be done slowly and with considerable care.

Once the issues or variables to be coded have been chosen the researcher may often be regarded as having taking a route from which

---

[6] Some researchers argue that there is no need to make any attempt to identify codes in advance of the coding exercise as they will simply "come to " the research as he or she proceeds with this exercise. However, a period of reflection is seldom wasted when it come to research.

deviation may be difficult. But this is not always the case as there can be considerable flexibility with regard to adding, or removing or combining codes.

With regard to establishing a list of codes it is useful if the researcher has a collaborator or a number of collaborators with whom he or she could work. If this is the case then each member of the research team could individually review the data and propose codes. These could then be reviewed and debated and a consensus arrived at. As coding data is a process of labelling concepts, the labels can always be changed if further reflection causes the researcher to take a different view of how to describe the data.

1.11 Data Coding

A central theme in Grounded Theory is that the researcher has to decide how the data could be grouped in order to deliver a higher level of meaning and this is typically done through the process of coding and subsequent manipulation and analysis of the codes. A code could be a single word or a group of words i.e. a phrase or even an acronym that will represent the concept being studied. Coding provides a level of abstraction whereby similar ideas are grouped on the basis that they represent a pattern or a theme and are thus assigned the same code. These codes may then in due course be assembled into high level concepts, categories or constructs.

In Table 1 seven issues are listed which have been mentioned by a senior manager during an interview and two possible ways of coding these issues raised are shown. Either of these approaches are suitable in this case.

18

Table 1: Example of open coding using a key word or an acronym

Issue	Code Example 1	Code Example 2
Our organisation is strong on strategic objectives	Strategy	STG
Our staff are rewarded on performance	Performance	PER
Clients are sought only when we feel that we have a solid value proposition to offer them	Strategy	STG
Staff conditions are above industry average in our organisation	Staff	STF
Our product development is closely integrated with our strategic thinking	Strategy	STG
Financial issues are dealt with by a separate team	Finance	FIN
We are very strong on staff self improvement	Staff	STF

When thought about, superficially coding appears to be a relatively unproblematic issue, but this is seldom the case. As well as the question of identifying the appropriate codes as illustrated in Table 1 above, there is the granularity of the codes. If the codes are at too high a level then some insights may be lost, but if they are too detailed the researcher may be subjected to data asphyxiation. Data asphyxiation is sometimes thought to be merely a pleasantry, i.e. something to smile about, but it can be a real problem and the researcher needs to be aware that collecting extraneous, nice to have, data normally only gets in the way. But when it comes to codes, it is of course easier to consolidate or group codes as the research proceeds than it is to recode data to a finer level of granularity.

In Grounded Theory coding is a step towards assembling concepts and can be demanding. Researchers can find that some data points could be coded in multiple ways and they struggle to decide which categorisation they should allow. Some data points are genuinely relevant to more than one code and when this occurs the researcher needs to allocate multiple codes. Too many multiple codes will dilute the effectiveness of the coding procedure.

Coding is often hard, tedious work which can take weeks to complete. It is important not to rush the coding process. Over an extended period of coding the researcher can lose sight of the objective. Coding is not an end in itself. In Grounded Theory, coding is a strategy for handling and organising large volumes of data in such a way that groups may be established which will lead to a better understanding of the data and the answering of the research question. The high level concepts, categories or constructs which are developed through the process of reflecting on the results of coding are building blocks of the theory which will be developed. Thus coding alone is only one step in the process of understanding, which will eventually lead to theory building.

1.12 Different approaches to coding

There are a number of different approaches to coding. In the first instance there is *in vivo* coding[7]. In vivo coding uses the exact words employed by the informants as the basis of the code. This is often regarded as an appropriate approach for any coding scheme. An alternative could be that the researcher uses synonyms rather than the exact words. The researcher may believe that this would provide a greater degree of scope or flexibility with the analysis which follows the coding. Some researchers believe that they should code to a predefined coding framework. In the Grounded Theory arena, researchers are primarily looking to create a theory and a theory may be envisaged as requiring input variables, an understanding of processes and one or more output. In this context a researcher may look for issues in the data which fall into one or more of these categories. To start coding with such a framework in mind will risk losing some of the detail that may be available in the data which can substantially enrich the research findings.

The codes described up to this point are often referred to as open codes. They may also be referred to as first cycle codes. Many Grounded Theory research projects will lend themselves to combining open codes into concepts of greater breadth and depth. This is referred to as axial coding or second cycle coding which involves creating higher level codes. These new concepts will normally be referred to as categories or constructs. It may even be possible that

---

[7] The name of this type of coding should not be confused with the software product of a similar name i.e. Nvivo.

categories or constructs could usefully be further combined into higher level categories or constructs and if this is the case an additional level of coding might be required. This has also been referred to as "lumping" ideas in order to create a higher level concept. Each level of coding should result in increasingly abstract concepts being articulated. However, at some point these labels or codes are finally grouped and it is these aggregated ideas which will become the building blocks of the theory which will be developed. This final stage of coding is sometimes referred to as theoretical or theory coding.

When coding is completed the codes are counted and grouped and sorted. There are a number of techniques available to assist the researcher in this respect. There are computer products which offer comprehensive tools and which facilitate the recoding and prioritising of codes. These are referred to as Computer-Assisted Qualitative Data Analysis Software (CAQDAS). In addition there are statistical processes available which can be used to create perceptual maps of the codes themselves or of the groups of issues which have been identified through the coding process. To produce perceptual maps[8] it is usually necessary to code both the issues being identified and also the type of individual who raised the issues. An example of codes which have arisen from different sources is displayed in Table 2.

---

[8] The term perceptual map is used here generically. There are a number of different ways of describing pictorial representations of ideas and concepts which include cognitive maps and mind maps. These usually have the same objective which is to show how a range of ideas may be seen in relation to one another.

Table 2: A two dimension table of codes suitable for perceptual mapping

Code	Total	Source 1	Source 2	Source 3
STG	35	20	10	5
PER	29	25	2	2
STA	25	20	3	2
STF	20	15	3	2
R&D	15	5	5	5
FIN	15	8	7	-
M&A	14	14	2	-

Coding is an important part of Grounded Theory, but it is only one way of understanding data. Data can be understood in a much more holistic way which does not require coding but which could comply with many or at least some of the other characteristics of Grounded Theory. The process of understanding data in a holistic way is often referred to as hermeneutics which requires a deeper understanding of interpretive research than is normally associated with Grounded Theory.

1.13 Second cycle coding and categories

The journey from a table of coded data to a theory requires considerable intellectual work and a creative mind set. Table 3 below shows some of the work required and may be described as follows.

Table 3: Some of the data acquired in moving from open coding to categories.

Issue	Open Code	Frequency	% Freq	Second coding for categories	Cum % Freq
Website development	WSD	28	9%	Pre-campaign	9%
Training for sales people	TSP	25	8%	Pre- campaign	18%
Produce design	PPD	23	8%	Pre- campaign	26%
Creation of after sales policy	ASP	22	7%	Pre- campaign	34%
Product descriptions	PPD	21	7%	Pre- campaign	41%
Returns policy	RRP	20	7%	Pre- campaign	48%
Packaging	PAK	18	6%	Pre- campaign	55%
Preparation of e-mailing lists	PEM	17	6%	Pre- campaign	61%
Warranty policy	WAR	13	4%	Pre- campaign	66%
Commission on sales	COS	27	9%	Finance	72%
Cost of after sales personnel recruitment	AFS	15	5%	Finance	77%
Credit terms	CRD	13	4%	Finance	81%
Finance to buyers	FIN	12	4%	Finance	85%
Debtors policy	DDP	11	4%	Finance	89%
Sales bonuses	SSB	16	5%	Evaluation	93%
Performance report-ing	PPR	12	4%	Evaluation	97%
Cash flow	CAS	10	3%	Evaluation	100%

The transcript has to be examined to determine what the most important issues are and each of these issues requires a code. In this example a three letter code has been used. The frequency of the occurrence of these issues in the transcript is counted and a percentage frequency is calculated as shown in Table 3.

The issues are then re-examined in order to establish if any of them could be understood as being part of a higher level set of issues. In this example the activities which are related to the original preparation of the product and the creation of the relevant marketing and sales policies are considered to be suitable for grouping under a heading of pre-campaign activities. In addition a number of the issues could be grouped as relating to finance and evaluation.

When these have been identified then it is appropriate to calculate the cumulative relative frequencies and to sort the table as shown in Table 3.

With this table the researcher can look for connections and patterns. It is important to emphasise that this is only a first step and that much more reflection than can be described here is required in the process of understanding, analysing and interpreting the data.

From Table 3 it can be seen that there are three major categories which emerge from this research. These are described as pre-campaign, which is mentioned 66% of the time by informants; then there is finance, which is mentioned 23% of the time and finally there is evaluation, which is mentioned 11% of the time. This analysis is then used as input into the processes that result in the emergence of the theory.

1.14 Research Memos

During the process of data collection the researcher is required to create Research Memos. A Research Memo is essentially a note which the researcher writes to himself or herself. It is common practice for all researchers to write such Memos. It is for this reason that researchers keep logs or diaries to which they can later refer. But in grounded theory there is a special emphasis placed on these memos which should be formally recorded and used in the process of understanding and analysing the data. And of course the Research Memo is itself a form of data just as a field note is when a researcher is involved in a series of interviews for example, and thus the Research Memos may also be processed by coding[9].

A Research Memo is in a sense a record of reflection which occurs to the researcher during the process of data collection or even when data analysis has been undertaken. But unlike fleeting reflections which often occur with little or no notice and which all researchers will experience from time to time, in this case the record is physically made and can be incorporated alongside either the data analysis or the final results of the research.

Sometimes a Research Memo can arise as a result of an entire research episode while on other occasions the researcher may feel it important to create a Research Memo as a result of a particular word

---

[9] The research memo will of course contain the researcher's opinions, interpretations and reflections as well as data. When using these, care has to be taken to try to be as unbiased as possible.

or a phrase used by an informant or when the data is being coded and some interesting relationship occurs to the researcher.

Research Memos are sometimes thought to be like the cement which keeps the building blocks of the research together. They often are used to bring context and to highlight relationships identified by the researcher to the other forms of data collection acquired through a formal data collection process.

A comprehensive set of Research Memos is often an indication that an adequate amount of reflection has been employed during the research. Table 4 is an example of a group of Research Memos which have been created in a short period of the research.

Table 4: Example of Research Memos

Date and time	Source or incident	Memo
June 2, 10.00	After meeting with Managing Director in Head Office.	Not convinced that she shares her vision for the organisation with the other members of the senior management team. Need to look again at the minutes of the board meeting for clarification.
June 5, 12.00	After meeting with the sales manager for the Northern Region in his office in Newcastle.	His attitude did not strike me as being collegiate. There is an organisational principle which states that the further one is away from Head Office the happier the "troops" are. I wonder if that is the ethos here? Must do some triangulation checking.

Date and time	Source or incident	Memo
June 9, 09.30	After the meeting with the marketing analysts.	Having had a round table discussion with 3 marketing analysts who are responsible for providing pricing advice to the Marketing Director it appears that they operate on an optimisation of transfer basis and this does not easily fit with the information received from the Head Office. Must check.
June 12, 16.30	While coding the data obtained during the interview with the Financial Director.	The relationship between the production manager and the head buyer as described by the Finance Director seems to be very cosy. Need to find out more about this.

1.15 Theoretical Sensitivity

Grounded Theory also addresses a concept called Theoretical Sensitivity. This is a particularly interesting concept as it goes directly to the heart of theory generation. Theory generation is regarded by all to be difficult. Firstly it requires both experience and imagination and a particular level of intellectual maturity combined with a degree of self critique. None of these attributes are simple or easily achieved. Some researchers will improve with experience but others will not. Different researchers have various ways of understanding the issue of Theoretical Sensitivity. Grounded Theory requires personal openness, an enquiring frame of mind and interpersonal skills on the part of the researcher. It has to be emphasised that when seeking theoretical explanations or understandings from data the researcher needs to set up a train of constant comparison to allow similar events and processes to be taken into account. This requires concentration skills. It is also important that the researcher remains self-conscious as to how his or her theoretical understanding is unfolding and be aware of any biases which might be creeping into his or her thinking.

To understand the original thinking of Glaser and Strauss it is important to be continually aware of the fact that Grounded Theory was created for the purpose of theory generation. The generation of a theory which purports to have any degree of originality is difficult. Our well established knowledge from our many years of education tends to intervene and influence whatever novelty we are aiming at. To try to minimise this it was proposed that the researcher should not delve too deeply into the extant literature. Many researchers have found this problematic. There are obvious challenges concerning the inherent difficulties when the literature has not been properly reviewed. But while this is certainly the case, the spirit of Grounded Theory is that the researcher needs to avoid the confirmatory bias which is so often a feature of academic research. This has always been an issue of considerable concern.

1.16 The Emergence of Theory

The aspect of research which requires the highest level of creativity is the development of the theory. Theory does not spring out of the data no matter how well it was acquired, coded, sorted and analysed. Neither does theory normally result from any mechanistic method to understanding the issues being studied. In Grounded Theory the researcher works with his or her codes to be able to acquire a fuller understanding of the properties and the dimensions of the categories and the concepts which have been derived from the data via these codes. This is achieved, inter alia, by comparison between different data sets and the researcher uses these to create clusters of categories and concepts while beginning to conceptualise relationships and thus create a route to possible theories.

It has to be admitted that some researchers will be better at this than others but even those who struggle with this at the outset should not abandon this activity as researchers will improve with experience as their Theoretical Sensitivity (see below) improves.

Theory mostly evolves or emerges when the researcher has carefully reflected on the meaning of the data and how it could answer the research question. It is intellectual processes or the cognitive capacity of the researcher which drive theory creation. But theory creation does not normally occur quickly and Grounded Theory recognises this and thus the researcher should be involved in persistent interaction with the data[10]. No timeframe can be suggested for this period of persistence but researchers should expect that they may be involved with this phase of the research for weeks or longer. It is well recognised that in qualitative research the process of data collection and data analysis are normally intertwined as the researcher's understanding cannot help but be influenced by what is heard and seen during the collection process. It is important that researchers are aware of this and that they develop a self critical attitude towards how they are interpreting the data they are collecting and indeed coding. There are always different perspectives and different points

---

[10] Grounded Theory requires an ongoing commitment to the research even when the researcher has taken time off to do other things. Flashes of insight can occur at any time.

of view and the researcher needs to create as many of these as possible and then recognise them as such and make an evaluation of them.

However, sometimes the researcher will not be able to achieve the insights required here on his or her own and will need one or more other individuals who chew over the data or evidence collected and what it might mean. Sometimes a dialectic process is required and while some experienced researchers can perform such a process on their own, those who are new to research can seldom achieve this.

Dialectic is the process of a thought being mulled over by more than one mind and some reservations, or additions, or objectives being raised. The first use of this dialectic process is attributed to Socrates and it has been an important platform for academic enquiry for a long time. It is often said that by using the form of the dialectic the best teaching is achieved, i.e. when the student and teacher are engaged in a debating process. This is because the teacher does not tell the students what to think, but rather allows them to work out for themselves what the matter being discussed really is about. In order to achieve the benefit of the dialectic, researchers are often encouraged to discuss their research with as many suitably informed individuals as they can find.

There is however a caveat regarding theory development. Sometimes theory does not appear to emerge from the data. This can be due to a number of reasons including the fact that inappropriate data was collected and/or an inadequate amount of data was acquired. More seriously perhaps is that the researcher cannot see the links or associations which were expected. This may be the result of inadequate analysis or it may be that the research question was not correctly conceived and expressed.

It is sometimes said that researchers force the data which suggests that a theory is evolved which has not really emerged. There is an expression that

"If you torture the data long enough, it will confess[11]"

and researchers need to be careful not to fall into such a trap.

It is important that the theory which emerges has a claim to original-ity. It is inevitable that in the Grounded Theory process, research is-sues and ideas which are already well known will be encountered. These are normally not of much value to academic research and the researcher needs to avoid the temptation of repeating 'mother-hoods' which will often be seen as platitudes.

It is because of the need to have an original theory emerge that Glaser and Strauss recommend that researchers should not spend much time acquiring an in-depth familiarity with the literature in the subject matter of the research. If the researcher has been thoroughly immersed in the literature then there is a higher probability that this already established knowledge will drive the researcher towards the

---

[11] The comment about torturing data has been attributed to Ronald Coase although it does not appear in any of his published works. Ronald Coase was an academic economist who was honoured with the 1991 Royal Swedish Academy of Sciences, "Nobel Prize" for Economics.

confirmatory bias and it is important for Grounded Theorists to avoid this. Of course for a number of important reasons it is not possible to engage in research without some knowledge of the literature, if for no other reason than the necessity to ensure that the research question has not already been addressed and answered by others. Perhaps it is more important to emphasise that it is simply not feasible for a researcher to approach any research question from a tabula rasa state. Previous experience and understandings are deeply rooted in everyone's cognitive capacity and it is not possible to disengage these fundamental parts of an individual's existence.

1.17 Substantive and Formal Theory

Grounded Theorists distinguish between Substantive and Formal Theory and an understanding of these terms is critical to the use of the method. Substantive Theory is that theory which has been grounded in the data acquired and for which there is clear evidence supplied by the study. Substantive theory will have a typical theoretical formulation and this will be recognised by some degree of abstraction. It may have a high degree of validity or in the language of qualitative researchers, plausibility. Like all qualitative research it is unlikely to be able to claim reliability in the sense that the precise research cannot be easily replicated. But if created competently substantive theory will have credibility. Furthermore the results of this type of research will only be able to claim a limited level of generalisability which could be referred to as transferability or perhaps trustworthiness.

Formal theory will have the same credentials but is recognised by a higher level of abstraction and it will have a greater claim to generalisation or transferability. It is recognised that before arriving at a point where a theory could be considered to be formal it will need to have previously been considered to be substantive.

It is the ideal objective of academic research to be regarded as having developed formal theory, but sometimes this is beyond the reach of a particular research project. Especially in social sciences, in most cases the creation of a substantive theory is adequate from the perspective of being awarded a degree or having a paper published in an academic journal.

1.18 Pragmatism - the Heart of Grounded Theory

There is another particularly critical issue which Grounded Theory reminds us of and this is to do with how the findings of academic research in the social sciences may be validated. Social science is driven by a problem-solving ethos. There are perhaps some aspects of social science where the term 'pure research' might be relevant, but most social science may be seen as a response to addressing a problem. When the researcher proposes a solution to the problem or research question, and if the research question has been adequately challenging, the social scientist can make an argument that he or she has added something of value to the body of knowledge. In academic research, from the point of view of being awarded a doctoral degree, this contribution has to be to the body of theoretical knowledge while at the same time having a demonstrable impact on practice. This in turn raises the question, *What is a suitable definition of knowledge?* There are many definitions of knowledge but the Grounded Theorist will normally choose the pragmatist definition which relates to the application of knowledge and what can be achieved with it.

In the context of Grounded Theory the idea that knowledge solves problems may be traced back to Charles Sanders Peirce and the Pragmatists. These were research philosophers working in the United State of America starting in the second half of the 19th century. The pragmatists believed that knowledge was a coping strategy for achieving human ends and as such if something constructed by humans worked for them then in some sense it was right[12]. This idea underpins Grounded Theory and points to the practical value of academic research in everyday working experience[13]. In the language of this approach the researcher needs to produce a theory which has both 'fit' and 'grab'. By 'fit' is meant that it will be obvious to all concerned that the theory which is developed can be seen to have emanated from the data that has been collected and analysed. If there is no direct connection between the data and the theory then it would be said that the criteria for claiming the theory had been created by the process of Grounded Theory is not valid. When it comes to 'grab' this term is used to denote the extent to which the community in which the research has been conducted accepts the theory. If the community feels that the theory does not deliver any practical value then it is quite probable that Grounded Theory has not been used in

---

[12] The pragmatists tended to use the word *true* in their arguments. Their definition of true was simply that if something worked it was true, by definition.

[13] This is largely what Kurt Lewin had in mind when he famously said, "There is nothing so practical as a good theory."

the way that it should[14]. Pragmatism is not uncontroversial. Some researchers want to understand knowledge as more than a coping strategy. They tend to regard knowledge as some reflection of truth. But the pragmatists did not use this word in any absolute sense. Of course there are numerous difficulties when it comes to the definition of the word "truth" and these are magnified if this word is not used in an absolute sense.

Pragmatism's popularity largely declined after the middle of the 20th century but it has reappeared in the form of the Neo-pragmatists. The leading figure in this movement was Richard Rorty. Neo-pragmatists will point out that the emphasis which universities place today on the need for academic research to have utility for the community it serves is nothing more that the articulation of the values of pragmatism and it is difficult to contradict this assertion. Thus, they would argue that whether a researcher admits to being a pragmatist or not, there is a substantial element of pragmatist thinking in the way that universities now see their remit. Some researchers feel that the concept of pragmatism borders on the simplistic. The argument they use is that even if the findings of the research works now, so what? They ask how long it will continue to work. In addition there are problems with the notion of something "working". How is that to be defined? What some people would regard as working well, others might say it was working poorly or even not working at all. Thus, the question arises as to whose opinion is to be accepted and how much

---

[14] Of course there could be other problems concerning the provenance of the research question.

credibility should it be given. From a philosophical point of view pragmatism is a quagmire. Despite this, pragmatism is a useful concept and offers an interesting way of discussing the validity of research findings, especially in the field of social science.

The concept of pragmatism is further explored in Chapter 5.

1.19 A Positivistic Mind Set

There has been a longstanding debate in the social sciences concerning the value of a positivist versus an interpretivist mind set as the principle driver of academic research activities. Definitions of positivism and interpretivism are not trivial as both of these definitions are based on a series of assumptions about research practice and the objectives of a research project.

Grounded Theory is sometimes described as requiring the application of a positivistic mind set. One of the reasons for this is that Grounded Theory relies on fracturing the data through the process of coding. Data coding implies that the data set acquired will not be understood in a holistic way, but rather by becoming familiar with the different messages embedded in the data. As coding is the first step in grouping data within the Grounded Theory method the researcher will tend to move the research along a non-holistic trajectory. Although some Interpretivists are comfortable with coding, others are not and they see this activity as being intrinsically positivist in nature. The opposite approach where the data is not fractured is referred to as hermeneutics, which requires the researcher to obtain an understanding of the text being studied without it being coded.

It is normally agreed that fracturing or parsing data can change the original meaning which the data should have conveyed and this is a challenge which anyone who attempts coding has to be able to manage. On the other hand some researchers argue that to capture the meaning of text, it is often essential to parse it.

The debate between the positivists and the Interpretivists, which can be acrimonious is misguided. Both of these research frameworks are perfectly valid. The question as to which route should be taken is often a matter of the research question and the type of evidence which is available. Increasingly academic researchers in the business and management arena are combining aspects of both these frameworks and employing Mixed Methods paradigms.

1.20 Different Schools of Grounded Theory

Some researchers suggest that there are three versions of Grounded Theory. The three most commonly named versions are the original Glaser and Strauss (1967), the Strauss and Corbin (1998) and the Charmaz (or lately the Charmaz and Bryant (2007)) versions. In fact there are as many different versions of Grounded Theory as those who employ this technique in their research. No two researchers will apply Grounded Theory in the same way. There is also the question of all those researchers who would not claim to be Grounded Theorists but whose thinking can be clearly seen to have been influenced by Glaser and Strauss. As mentioned earlier, Grounded Theory has since its original articulation made an important impact on social science research thinking and practice in a much broader sense than would have been anticipated of a particular research method. From this alone one can see the importance of Grounded Theory.

1.21 Academic Credibility

Academic Credibility is based on two issues, which are whether the research may be regarded as having added something of value to the body of knowledge and the quality of the scholarship employed in so doing.

Grounded Theory provides a framework with which it is possible to argue and to justify the argument that knowledge has been created. In this respect Grounded Theory is attractive to academic researchers and this accounts for the substantial growth in its popularity during recent years. As mentioned before, Grounded Theory opened up what was the "black box" of qualitative research and allowed the academic community to see and understand the "nitty gritty" involved.

However, the issue of scholarship is different. Academic scholarship has a number of attributes or dimensions. The first of these relates to the fact that a scholar has to be well read to the point that he or she is aware of the thinking of all the authorities in his or her field of study. The second dimension of scholarship is that the scholar is required to be able to craft a convincing argument concerning all aspects of the research conducted.

In its original form Grounded Theory called for researchers to avoid too much contact with the extant literature for the reason that knowledge of it would inevitably lead the researcher into a theory test mode rather than a theory generation mode. Even when this did not happen, being too close to the literature could be expected to induce or strengthen a confirmatory bias which is present in much research. There is an interesting comment attributed to Darwin (Darwin 1978) concerning this issue which is:

> *In many ways it was fortunate indeed that I was not a fully trained naturalist or geologist. My mind was not yet enclosed within the confines of a single discipline but was free to wander and speculate on many things.*

Knowledge is a positive attribute for a researcher provided that it does not in any way reduce the researcher's ability to look at new situations with an open mind.

It is probably the case that in the 21st century academic research is more outward looking and broader in its scope that 19th century science but nonetheless Darwin's point has some validity. The Grounded Theorist has to be careful about the influence of the extant literature, while at the same time demonstrating that he or she is aware of the thinking of the authorities in his or her field of study.

With regards to scholarly argument crafting the same standards apply to Grounded Theory research as to any other academic output.

1.22 Summary and Conclusion

Grounded Theory is not the only method by which theory may be generated, but it has become one which has earned a considerable amount of respect in the social science community and which all researchers should be aware of, even if they themselves decide not to be a Grounded Theory practitioner.

Grounded Theory offers a comprehensive array of both intellectual tools and specific practical advice on how to conduct rigorous research which will be recognised as having made a contribution to the body of theoretical knowledge. The fact that there has been a schism between the originators of Grounded Theory does not in any way detract from its value. However this schism has created a concern among new researchers who tend to believe that to benefit from the use of Grounded Theory a research paradigm has to be followed precisely and of course this is not necessarily the case. Quality theory generation can be achieved following either the so called "classical" routes described by Glaser and Strauss or the revisionists, or researchers may use the different intellectual components and tools described in this Chapter to find their own way to theory creation. In short, academic researchers can avail themselves of those parts of the Grounded Theory method which they believe will enhance their work. In such cases they may describe their approach as *"having been informed by Grounded Theory"* or *"being Grounded Theory like"* or *"being in the tradition of GT"*.

But Grounded Theory is not without its detractors. Critics of Grounded Theory complain about how much work is necessary to comply with the approach that has been spelt out in the original texts. There is little doubt that if the method articulated in the Grounded Theory texts is followed and no short cuts are taken then these processes are time-consuming. But academic research has to be undertaken with considerable care and reflection and is therefore intrinsically time-consuming.

There may be just too much work involved with the Grounded Theory method for the academic researcher. By focusing on the detailed work required, researchers can get lost in the trees and forget that they are there to study a plantation or woods. One of the most amus-

41

ing interpretations of this detail was supplied by Alvesson and Skold-berg (2009) when they said:

"Grounded theory is a doctrine for those who have strong re-ligious meanings particularly of a Protestant persuasion[15]. Success entails a process of purification, achieved by aban-doning theoretical blinkers. Only then can diligent scholars see reality as it is, only then can they see the light. This blessed state is obtained through hard conscientious toil. Only the truly industrious – those who have collected an enormous amount of data and have lifted them up into various cate-gory-containers, which in turn are stacked up to create a great cathedral-like structure, on top of which sits the theory (with a small 't') – can avoid the corruption and perils of free thinking and hedonistic theoretical laxity, which allow the un-godly to think great, free and dangerous thoughts, abandon-ing the straight and narrow but rightful empirical path. Grounded theory is the path for petite bourgeois Protestants, where the accumulation of data leads to modest but certain returns in the form of limited but secure stock of cultural capi-tal. The grounded theorist is Luther's man".

---

[15] The term Protestant persuasion alludes to the Protestant Ethic which was an ex-pression coined by Max Weber who argued that spiritual salvation and worldly pros-perity could be achieved by the work and thrift ethic. It was for this reason, it was argued that on average those living in Protestant orientated societies enjoyed a higher standard of living than the average person in non-Protestant societies.

Of course these comments are not to be taken too seriously but neither are they entirely in jest. The Grounded Theory method is not for the faint hearted or for those who want a quick solution to justifying a piecemeal methodological approach.

It is important to be aware that some researchers believe that to refer to a research project as being based on Grounded Theory, all the techniques described by either Glaser and Strauss or Corbin and Strauss should be followed rigorously with no omissions. But this is probably a minority view.

It has been argued that Grounded Theory is more suited to group research or joint research projects than individual research in that the use of a group will facilitate establishing robust codes and a group approach will probably lead to richer insights when exploring the intricateness of the data. When theory is being created the existence of a group implies a greater facility for the employment of the dialectic. But Grounded Theory is regularly used in research conducted for academic degrees and not many universities allow joint research projects for this purpose. In such cases the researcher may call upon his or her supervisor for this type of assistance or work with other research degree candidates on a reciprocity basis.

Grounded Theory requires a high degree of maturity which includes patience with the theory emergence process as well as the ability to be self critical. For this reason alone Grounded Theory is not to be recommended to all researchers at all levels.

Grounded Theory is something which all social science researchers should be aware of and some researchers may put into practice.

2

The Trouble with Grounded Theory?

Although methodology is empowering it can also be highly restrictive when creativity is required.

2.1 Methodology everywhere

Too much is sometimes expected of research methodology and research methods. It is sometimes thought that there is a particular way to do research and to discover new knowledge, and that if only we could find this way we would be able to produce good research and have truly impressive findings. Alas, this of course is not true and the fallacy of this idea has been known since ancient times. Aristotle is reputed to have said, "There is no method other than intelligence". A more modern version of this idea was expressed by Kaplan (cited by Rosenthal and Rosnow, 1991 p.27): *"One contemporary philosopher, Abraham Kaplan, when asked to define the scientific method, answered that the 'scientist has no other method than doing his damndest'."*

So what is the basis of the extraordinary amount of energy expended on teaching and examining research methodology? Why do we have maybe, several thousand methodology and method books to choose from? There is no short answer to this question. Are we over-sensitive to the possibility of our research being challenged on the grounds of validity, reliability, generalisability, to mention only three possible areas of weakness? The answer to this is certainly, Yes, and to explore the reasons for this a little historical context is useful.

2.2   Methodology as a modern concept

There are a few important issues to understand in making sense of this largely excessive focus on research methodology. Modern research came about in opposition to the previous way in which humans knew about the world, which was through revelation. Before modern science we were told about the world through the holy books, being informed about them by sages, normally spiritual leaders, who spent their lives studying these texts. The key issues were belief and faith. Men and women who were not privileged were not entitled to question this received wisdom and the penalty for so doing was unthinkable. After all, questioning the Word of God had to be a serious sin and thus a heinous crime. Giordano Bruno was the last man to be burnt at the stake for being so impertinent; they called this type of sin, heresy – meaning to question the established knowledge of the Church. Science was born in an atmosphere of fear and thus scientists had to be most careful about what they said and how they said it. Some science authors suggest that Copernicus did not attempt to publish his book for fear of his thesis that the earth travelled around the Sun being regarded as heresy and thus he would possibly face the Inquisition. Copernicus died before the impact of his ideas was felt[1]. Galileo, who was a man less cautious and prepared to

---

[1] There is a sort of parallel here with Darwin who did not publish his work because of the possibility of his ideas giving offence. This was of course social pressure. And

play fast and loose with the authorities in Rome, paid the huge penalty of having to spend 27 years under house arrest, but at least he was not executed as happened to Giordano Bruno[2].

Science during the Enlightenment period generally did better in the Protestant countries than those dominated by Roman Catholicism. By its very nature Protestantism is less centralised and thus generally there was less formal control exerted over the scientific community. This freedom was used by scientists and philosophers to explore issues and arrive at conclusions which would have been problematic in Roman Catholic countries. The most famous philosopher to seek refuge in Protestant Europe was Rene Descartes who left France for fear of Papal intervention.

The intellectual freedom enjoyed by Protestant researchers did not mean that they were any less devout in their religious beliefs. After all atheism or even agnosticism were not considered as respectable alternatives to Catholicism until much later.

So as modern science developed it became understood that it was essential that a *cast-iron* case was made for any of the knowledge

---

indeed when he did publish the concepts in the book *The Evolution of the Species*, they did give great offence to some. Of course Darwin did not face being brought before the Inquisition but nonetheless he was scared of how his work might be received and he knew his ideas would offend his wife.

[2] Galileo was convicted of heresy and was forced to recant before being placed under house arrest. It was only when Pope John Paul II apologised in the year 2000 for mistakes made by Rome that an apology was made for the way Galileo was treated.

claims scientists made[3]. It was for this reason that empirical research became the primary modus operandi in the 17th and 18th centuries. This empirical research focused on experimentation and the use of mathematics wherever possible to explain what was being studied and what findings were being claimed. This approach was eminently sensible and vast amounts of knowledge were developed using this experiment together with mathematics wherever possible in labora-tory or laboratory-like conditions, and so an understanding of how science should be conducted emerged.

2.3 No universal codification of the scientific method

As science further developed into the 19th and 20th centuries, so did the idea of what it meant to conduct research in a way which was acceptable to the international scientific community. Although there was no universal codification of the rules of research, the scientific tradition, now generally referred to as the scientific method, was passed on from one generation of scientists to another. In many sci-entific circles the scientific method was elevated to the status of be-ing the only way of conducting research. There is little doubt that it was useful to have the scientific method as an amulet[4] as long as it

---

[3] The fallibility of research, although debated in some scientific and philosophical circles, was not a real issue until later. It was only when Karl Popper developed the notion of falsifiability that scientists began to really understand the contingent na-ture of knowledge.

[4] The word amulet is used here in its broadest sense which suggests that following the scientific method prevents scientists from taking a wrong approach and thus going astray. In research there is always the danger of error and the result of the research being regarded as pseudoscience or worse

was accepted as being like the entities in Plato's Cave Parable i.e. an "ideal form". In reality there were many occasions in which the ideal of the scientific method was not rigorously applied and yet *adequate scientific results* [5]were achieved. But an unrelenting belief in the scientific method still persisted in many quarters of academe.

It also became clear that the dangers of the late medieval attitude to knowledge creating had abated as Christianity splintered and the influence of the Church contracted from much of civil life.

William Whewell (1996), one of the early critics of the scientific method, declared:

> *An art of discovery is not possible. We can give no rules for the pursuit of truth which should be universal and peremptorily applicable.*

Yet the notion of the quintessential importance of the method to academic research remained. Despite evidence that there were other routes to scientific enquiry, scientists could not let go of the tradition of insisting on the supreme importance of the method. The dilemma was that if it were true that it is not possible to give *rules for the pursuit of knowledge* then we were admitting to not having any control over how true knowledge could be created. How could we then distinguish between knowledge and pseudo-knowledge?

---

[5] An *adequate scientific result* is one which satisfies the scientific community.

2.4   Anything goes?

But much worse was ahead for those researchers who believed that it was necessary to have a strict formula for research practice and a severe body blow was dealt to the believers of the scientific method by Feyerabend (1990) who pointed out :-

> *It is clear, then, that the idea of a fixed method, or of a fixed theory of rationality, rests on too naïve a view of man and his social surroundings. To those that look at the rich material provided by history, and are not intent on impoverishing it in order to please their lower instincts, their craving for intellectual security in the form of clarity, precision, 'objectivity', 'truth' it will become clear that there is only one principle that can be defended under all circumstances and in all stages of human development. It is the principle: anything goes.*

Feyerabend's term *anything goes* is sometimes misunderstood. It can be fallaciously thought that *anything goes* means that there are no scientific standards which of course is not the case. What Feyerabend's insight was is that there is no point in believing in a prescribed path or route for research. There are many ways to research and each researcher has to find his or her own way of answering the research question. What decides if the research should be respected are the findings and what they actually mean to the scientific community. And the decision as to whether the findings are adequate is made by the corpus of scientific researchers in the community. The scientific community is sometimes referred to as the Republic of Science[6].

In terms of research degrees, especially a doctorate, this means that the researcher may follow a highly structured method throughout his or her research or he or she may choose a much more open and emergent approach to the research. What matters, i.e. what makes the research credible, is not simply the methodology, but the findings and whether they are credible to the supervisor(s) and the examiners[7]. Of course in academe the researcher has to be scholarly and it has become customary, at least in some cases, to demonstrate a scholarly knowledge of research methods as well as scholarship in the subject content.

It is important to realise that findings alone, no matter how interesting they are and how beneficial to the community, need to be supported by scholarship if an academic award is being sought.

2.5 Theorising and theory development

Academic research distinguishes between theorising and theory development. Theorising involves informal postulating of possible theories. Sometimes this is useful from the point of view of thought experiments, but mostly it is considered to be too superficial to be taken seriously for the purposes of academic research. Theory devel-

---

[6] The Republic of Science: Its Political and Economic Theory by Michael Polanyi, http://www.missouriwestern.edu/orgs/polanyi/mp-repsc.htm
[7] Although different methodologies may be chosen and a researcher may even produce his or her own methodology, a research degree requires a high level of scholarship and this acts as a break on facetious methods being proposed.

opment on the other hand requires a deeper level and more serious reflection. Grounded theory offers the modus operandi to achieve theory that will be valued by the academic community. Thus although Feyerabend is logically correct many researchers prefer the comfort of a method such as Grounded Theory.

2.6   Prevention from self deceit

But does this mean that there is no need for research methodology or for it to be taught and examined? The answer is certainly, No. Research methodology is imperative for two reasons. First of all it provides a language which facilitates researchers' communications about their ideas, without which there would be very little prospect of their being able to understand one another. Secondly it remains a type of compass, but not for checking that they are on the one right path to discovery but that they do not descend into self-deceit. Raimond (1993) expressed this well when he said:-

> *Scientific methodology needs to be seen for what it truly is, a way of preventing me from deceiving myself in regard to my creatively formed subjective hunches which have developed out of the relationship between me and my material.*

It is far too easy for researchers to convince themselves that sound research practice is the basis of their study, while a variety of inappropriate attitudes, biases and activities are being employed.

Novice researchers inevitably need a prescribed and detailed methodology. They need some sort of way of focusing their creativity and a methodology such as Grounded Theory can be useful in this respect, especially if it is used as a guide. A new researcher can find the notion of an emergent approach to their research daunting. Some newcomers to research even want to be told how to do their research and they sometimes ask for detailed instructions to be given to them on a weekly basis. Some, though not all, researchers perceive Grounded Theory to play an important role in this respect. They believe that to be a Grounded Theorist it is necessary to follow a highly prescriptive path which is broken down into distinct activities such as collecting data, coding data, sorting data, re-coding, memoing etc. There is also comfort in coming to terms with the vocabulary of Grounded Theory. But this comfort can also involve self deceit. Learning what Grounded Theory is and knowing the framework and the steps required does not mean doing good research. It is sometimes suggested that following a detailed research method is like painting by numbers which of course can produce a competent looking piece of work, but cannot deliver a piece of work which reflects anything about the person of the artist.

Other researchers understand that Grounded Theory can deliver a mindset which enables academic research to be conducted in a much more insightful way without having to address all the steps prescribed.

2.7 Methodology and creativity

From a more general point of view excessive methodology can have an adverse impact on creativity. The act of knowledge creation is essentially creative and it draws heavily on the imagination[8] of the re-searcher. Without this characteristic being present in abundance re-searchers can find it hard to know what to do with their data and even their research findings, and their projects can simply collapse[9]. Richard Feynman (1995) made this point when he said:-

> *But what is the source of knowledge? Where do the laws that are to be tested come from? .... But also needed is imagination to create from these hints the great generalizations – to guess at the wonderful, sim-ple, but very strange patterns beneath them all, and then to experiment to check again whether we made the right guess.*

But too much methodology can smother the imagination required for good research. Researchers can become so preoccupied with follow-ing the rules of the method that they lose sight of the essential na-ture of research, which is to try to see what they are studying through new eyes. Furthermore, no amount of method can compen-sate for this. In fact the more the research focuses on method and methodology the more likely it is that imagination has taken a back seat. This is at the heart of the lament voiced by Alvesson and Skold-berg (2009) which suggests that Grounded Theory is of interest to those of the Protestant Persuasion.

---

[8] This is driven by experience, maturity and intuition.
[9] It is quite correct to say that without any methodological education or experience a researcher will not know where to start. What is required is a balance where methodology is not seen as a set of restricting rules.

It is worth pointing out that the comments of Richard Feynman (1995) above were made in the context of physics. In social science we generally do not have the same access to experiments that physicists have and thus we generally cannot easily experiment. In the social science world, testing often takes the form of obtaining more observations or other evidence and seeking confirmation that this evidence offers some support for the theory which has been articulated.

For novices, Grounded Theory may be seen as the answer to the *"prayers"* of those who want a method. So much is spelt out in detail. There are many steps to be followed and definitions provided. But the more experienced the researcher becomes, the more likely he or she will want to explore his or her own ideas of knowledge discovery and perhaps they will feel that they could describe their work as having been informed by Grounded Theory, or maybe they will want to claim that they have developed their own approach to answering their research question.

Let researchers beware. Too much method which is not accompanied by imagination can be dangerous to your research.

For the more experienced researcher Grounded Theory takes the lid off the "black box" and allows the research activity to be explored and understood. Then it is up to the researcher to decide how much of the method he or she wishes to incorporate in his or her own research.

2.8 Summary and conclusion

Science is no longer obliged to the scientific method as the only way to create knowledge. Scientists may now take a much broader view as to what constitutes "good research". This does not mean that there is no pseudoscience being presented, but scientists are quick to alert society to such matters. The value of science and the quality of scientific endeavour is now measured by the scientific community itself.

The relationship between science and religion has been an interesting one, with the notion sometimes expressed that science has banished religious beliefs especially from the minds of intellectuals, being over stated. Many of the greatest scientists have had deeply held religious convictions. Einstein's famous comment, "God does not play dice with the universe" is but one example of a deeply rooted religious belief. What scientists have achieved is that they have reserved the right not to be told by religion or any other authority *how the world works*. The answer to this question is only acceptable if it has been arrived at by scientific endeavours.

To answer the question posed by the title of this chapter, there is no trouble with Grounded Theory. For novices to research it is helpful and a comfort to have a prescribed route to quality academic research. For the more experienced researcher Grounded Theory enhances our understanding of research and also acts as a reminder of what academic research is about and it furnishes us with a vocabulary with which to discuss important research issues.

However it is worth remembering that Grounded Theory should be used in the way that is most appropriate to the task at hand and not followed blindly as if it were a recipe.

3
Theory and Theoretical Research

*The raison d'etre of academic research has been the creation of theo-
retical knowledge?*

Grounded Theory involves empirical research using an inductive approach and leading to theory generation or development. It is therefore essential that anyone considering the use of Grounded Theory
has a clear understanding of the nature of theory and how it is developed in academe. One of the essential requirements of academic
research at doctoral level is that the research degree candidate can
point to how he or she has been able to add something of value to
the body of theoretical knowledge. This new theoretical knowledge is
sometimes referred to as the contribution of the research and without a contribution of some substance the research will not be considered adequate at doctoral level.

One of the striking characteristics of the Grounded Theory discussion
is the lack of discourse of what theory actually is and to many this is
not really surprising because of the confusion surrounding the word
"theory" which is used in a number of different ways. Although it is
perfectly correct, from an English language point of view to say, "My
theory is that John did not travel with the team to France because he
is frightened of being sea sick" or "The school teacher's theory was

59

that Jill had not spent enough time on her home work", what is being referred to here as theory is not *a theory* in the academic sense of the word. Even a quick review of the word theory suggests that there are many different meanings that may be attributed to this word "theory".

In addition to discussing the nature of theory from an academic point of view it is useful to contrast theoretical research with empirical research and thus this chapter addresses the processes involved in this other research paradigm.

Of course Theory and Theoretical Research are very broad subjects which could be regarded as worthy of a full book on their own. But this Chapter provides a starting point for understanding the issues involved.

Like the rest of this book this chapter/paper addresses a more general audience than those only interested in Grounded Theory.

This chapter/paper is based on work which was originally co-authored as an occasional paper by Dan Remenyi and Arthur Money.

Theory and Theoretical Research

The owl was the wisest of animals. A centipede with 99 sore feet came to him seeking advice. 'Walk for two weeks one inch above the ground; the air under your feet and the lack of pressure will cure you,' said the owl. 'How am I to do that?' asked the centipede. 'I have solved your conceptual problem, do not bother me with the trivia concerning implementation,' replied the owl.

(Shubik, 1984: 615)

3.1 Introduction

This paper discusses an approach to conducting research that relies on a non-empirical paradigm, which we will refer to as theoretical research[1]. All academic research[2] requires strong theoretical under-

---

[1] Perhaps the single most important difference between theoretical research and empirical research is the fact that the theoretical researcher does not directly collect primary data or evidence. The theoretical researcher does not conduct experiments or collect data through questionnaires. The nearest a theoretical researcher may come to this is to employ the idea of thought experiments as performed by Einstein, which involves the application of imagination and creative thinking to a hypothetical situation.

[2] The range of academic research usually includes work conducted for a masters or doctoral degree, research conducted for the purposes of publishing in a peer reviewed journal or work produced during a post-doctoral appointment. Of course research conducted by a university in terms of a commercial contract would not

pinning of some form. This is especially relevant to doctoral research, on which this paper focuses. For this reason we have used the term non-empirical or theoretical to describe research which predominately relies on an examination of the literature, reflection and discourse with knowledgeable members of the appropriate academic community. The creation, development and application of theory is the backbone of academic activities. Without theory academic activity would not exist in the form that we know it. Theory informs not only academe, but also the application of ideas in each field of study. It is therefore of paramount importance to understand the nature of academic theory and its purpose or role[3].

3.2 Theory underpins academic thinking

Academics attempt to work in general conceptual structures that offer a wide understanding of the concepts in the discipline. For this reason academic thinking involves rooting knowledge in theoretical frameworks. Without a theoretical framework, knowledge can only be quite specific and although such knowledge can be useful in a particular set of circumstances it may have restricted application.

---

usually be regarded as academic research. The principles discussed in this paper apply mostly to doctoral research. However they are also relevant to other academic research endeavours.

[3] Views of what theory is and how it works can be quite negative. It is frequently announced in a pejorative way that something might work in theory but not in practice. On the other hand Kurt Lewin contradicted this when he said, 'Nothing is as practical as a good theory', http://www.phrases.org.uk/bulletin_board/10/messages/290.html, viewed on 7 August 2013. John Maynard Keynes (1936) famously made a similar point. Garret Fitzgerald, the former Taoiseach of Ireland, is said to have exclaimed at a suggestion made to him that 'This may well work in practice but does it work in theory?'

Theory is therefore an integral part of any academic study and even in introductory university courses the teaching of theory will play an important role. In such courses theory will often be expressed as laws, principles or theorems. An introduction to Economics may address the Law of Diminishing Marginal Returns or the Law of Demand and Supply. In Physics it may be Archimedes Principle, in Chemistry it may be Arrhenius' Theory and in Mathematics it may the Binomial Theorem[4].

Little time or attention is normally given to defining the meaning of the concept of a theory or describing its nature, or even looking at what sort of limitations it may have. The term 'theory' is so deeply rooted in everyday language that it is assumed that its meaning is clear. But, like many other assumptions in academe, this is not always the case, as the notion of theory is neither obvious nor simple. When dictionaries are consulted it is revealed that there are a number of different meanings associated with the word theory and this certainly can be confusing.

3.3   Different meanings of the word 'theory'

Of the various different meanings of the word 'theory', listed here are a few that demonstrate the wide range of situations in which the word is used:

---

[4] Distinguishing between different types of theory, it is said that there is Normative theory – what to do; Descriptive theory – how things are; Analytic theory – why things are the way things are, how things work; and Critical theory – how things should be. In addition in the field of business and management research there is the question of Mode 1 and Mode 2 research/theory.

- A theory is a speculation – my theory is that the first horse to lead the pack in the race will very seldom finish first.
- A theory is a belief – my theory is that if you spare the rod you will spoil the child.
- A theory is a guess – my theory is that he will come later this afternoon.
- A theory is abstract reasoning – my theory demonstrates how value is associated with the perception of utility.
- A theory is a series of inter related concepts – the concept of a black hole brings together a number of different astronomical issues and relationships.
- A theory is an explanation – Newton's Third Law of Action and Reaction explains why motor vehicles are wrecked when they collide at speed.
- A theory is an aid to comprehension – Einstein's Theory of Relativity allows us to better understand how time and space are interconnected.
- A theory is a component of a body of knowledge – Modern Architectural Theory rejects the linear structures fashionable in the 1960s.

This range of usage of the word theory has not been helpful to Grounded Theorists or for that matter other academic researchers. As Grounded Theory is a theory developing method the importance of a clear view of the nature of theory is of utmost importance.

Theory is an example of one of those words of which one has to be careful if its meaning is to be understood.

It is the last five usages listed above of the word theory that are primarily used by academics. In our view a definition of theory is that:

A theory is systematically organised knowledge applicable in a relatively wide variety of circumstances, using a system of assumptions, accepted principles and rules of pro-

cedure devised to analyse, predict or otherwise explain the nature or behaviour of a specified set of phenomena. But it is also often simply the best explanation which is available at that time.

It is sometimes argued that theory needs to deliver some degree of utility or practical application. In this respect a theory is often looked to as a means of being able to predict and thereby giving the user of the theory some degree of control.

Within the original definition Grounded Theory induction is mentioned. But at the same time Abduction is introduced. It is probably fair to consider Abduction as a subset of induction although there are some researchers who strongly dislike the whole notion of Abduction.

Predictability is not a *sine qua non* and there is much useful theory which expands understanding by explaining, but which is not capable of, or simply does not lend itself to, prediction. The Darwinian Theory of Evolution is regarded by some to be an example of this (Darwin, 1986).

Theory may be developed by the process of induction or deduction. Whether induction or deduction is the chosen stratagem is really a function of the type of research question. In general, deduction would be used when there is an established theory to explore further and induction is used when a new theory is being developed (Remenyi et al., 1998). Deduction may be described as moving from a general concept to a specific situation, while induction involves moving from a specific situation to a more general principle. This is well discussed by Alvesson and Sköldberg (2001).

3.4   The limits of theoretical understanding

It is not necessary for a theory to be a complete explanation of a phenomenon. In fact many theories are only partial explanations. The full complexity of the situation may not be apparent when the theory is being developed. Even when a theory appears to be comprehensive it may not be able to encompass all the issues in its immediate domain. This point has been made by several philosophers of science, including Feyerabend (1993: 39) who said:

> *We may start by pointing out that no single theory ever agrees with all the known facts in its domain. And the trouble is not created by rumours, or by the results of sloppy procedure. It is by experiment and measurement of the highest precision and reliability.*

Feynman (1995: 2), describing the lack of comprehensiveness of our understating or knowledge, points out:

> *Each piece, or part, of the whole of nature is always merely an approximation to the complete truth, or the complete truth so far as we know it. In fact, everything we know is only some kind of approximation, because we know that we do not know all the laws as yet. Therefore, things must be learned only to be unlearned again or, more likely, to be corrected.*

In effect, at any given moment our theoretical understanding of any subject is always contingent on our current level of thinking or our current cognitive capacity. Our current level of thinking is nearly always in a state of transformation or development. It is often a struggle to articulate our current best understanding and any given state of knowledge needs to be thought of as an interim position. New ideas and new developments can arrive at any time and these can profoundly change our view of the world or even the universe.

Checkland (1986) provided an elegant statement of the lack of finality in respect of our understanding of the world when he pointed out:

> *Obviously the work is not finished, and can never be finished. There are no absolute positions to be reached in the attempt by men to understand the world in which they find themselves: new experience may in the future refute present conjectures. So the work itself must be regarded as an on-going system of a particular kind: A learning system which will continue to develop ideas, to test them out in practice, and to learn from the experience gained.*

Thus in the majority of cases theory provides useful but limited explanations and understandings of the world. It is important to realise that by their very nature theories and theoretical assumptions and explanations are likely to change continually. Nothing is ever fully settled for any length of time.

3.5   Research Paradigms

There are a number of different research frameworks that every researcher needs to be aware of and from which a specific strategy should be chosen. Two of the more important approaches are empirical or theoretical research. As mentioned previously empirical research, which is by far the most common, draws on experience or primary evidence in order to understand a phenomenon. Here the research question is studied by means of direct observation, accounts of phenomena recalled by informants or experiment. Empiricism may be described as an approach to research which postulates that all knowledge comes from, and must be tested by, sense experience (Locke 1974). Empirical research requires the evidence collected to be analysed and then synthesised, which leads to adding something of value to the body of knowledge. There are many ways of conducting empirical research and so it may be positivistic or it may be interpretivistic, to mention two major approaches.

> *Grounded Theory is always empirical and because it is a theory discovery method it is by definition deductive in nature. It is important for researchers to be aware of the lack of logic justification for deduction. This curious fact does not in any way reduce the importance of deduction. Some social science researchers consider the arguments against deduction to be spurious.*

Enthusiastic support for empiricism leads in its extreme form to the conclusion that, not only has all knowledge to start with the observation of experience, but also that empiricism cannot take the researcher beyond actual experience. In short this means that it is not possible to know anything other than that which comes directly or indirectly from observation.

This position is sometimes referred to as a form of scepticism, according to which, claims about subatomic particles or physical forces are just as doubtful as claims about supernatural entities. Those who hold this position would argue that science has to be seen as the discovery of relationships between the observable phenomena. This is a limiting point of view, which is not subscribed to by many scientists.

However it should be noted that, from an academic point of view, empirical research needs to be conducted within a theoretical framework and needs to have as its objective the addition of something of value to the body of theoretical knowledge.

3.6   Non-empirical or theoretical research

In contrast to empirical research, there is theoretical research, which is often cited as an important methodological strategy in business and management studies. The following is our understanding of theoretical research.

Theoretical research involves drawing on established ideas and concepts from published and non-published sources, especially the literature, and through a process of reflection and discourse develops, extends or in some other way qualifies the previous work to create new explanations, insights and theories, which provides better or fuller explanations of the issues and the relationships being studied.

Through theoretical research, it is possible to make a considerable contribution to the body of knowledge without having to collect or analyse primary data or evidence. For theoretical research data may of course be used by reference to already-published sources, which will thus be, by definition, secondary data.

In the business and management studies world, theoretical research is not always well received. In fact some academic researchers would argue that the process described above as theoretical research should not be regarded as 'proper' academic research. The basis of such a claim is that this type of theoretical research does not have a test component. This implies that theories cannot be postulated without any 'proof' or confirmation. For this type of researcher, 'proof' needs to be empirical and without the activity of 'proving' theory, it does not have academic status. In some cases this is referred to as theorising and is not seen as rigorous academic research. However this type of thinking is a misunderstanding of the nature of research. All research processes require conceptualisation. It is the starting point without which research and especially academic research cannot take place. The cerebral nature of research is well demonstrated by reference to Ashall (1996) who pointed out that:

> *Once, when asked by someone if they could see his laboratory, Einstein took a fountain pen from his pocket and said, 'There it is!' On another occasion he commented that his most important piece of scientific equipment was his wastepaper basket where he threw much of his paper work containing mathematical computations.*

Clearly Einstein was talking about theoretical research. Equipment or large volumes of data are not a prerequisite for theoretical research. Theory is created in the mind and this is perfectly respectable academic research. In a sense theoretical research is the modern equivalent of Rationalism. Thus rationalism is a philosophical view. It regards reason as the primary source of understanding. Believing that reality has an inherently logical structure, rationalists assert that a class of knowledge exists that the intellect can understand (Honderich 1995).

There were various schools of Rationalism including the Continental School begun by Descartes and the British School of Empiricism, which is said to have begun with the work of Locke[5]. Rationalism or theoretical research has long been the rival of Empiricism. In contrast to Empiricism, Rationalism holds reason to be a faculty that can access truths beyond the reach of sense perception, both in certainty and generality. The roots of Rationalism can be traced much further back in history than René Descartes. In his *Republic*, Plato points out that empirical evidence is problematic and he uses the notion of ideal forms to point out some of the difficulties with empiricism. In turn Galileo, although he was primarily an empiricist, takes this point further and moved science distinctly beyond the observational. Thus, even in the seventeenth century it was already becoming established that observations alone cannot supply an entirely satisfactory explanation of the physical world. This is not to say that science in most instances did not and still does not rely heavily on observations (Feynman, 1995).

Another aspect of theoretical research important to consider is that the material with which the theorist works need not be especially original. It is of course correct that doctoral research is required to demonstrate originality. A paper will not be accepted by a peer-reviewed journal unless it has something new to say.

---

[5] Modern empiricism is regarded as having begun with John Locke (1632–1704) with his clear attack on metaphysics in his essay 'Concerning Human Understanding', published in 1690.

Although Grounded Theory demands the condition of originality the distinction between substantive and formal theory is useful. Researchers not using Grounded Theory would tend to see themselves as either producing formal theory or aspiring to the production of formal theory. Grounded Theorists can stand back a step and lay claim to an equally important but perhaps a little less elusive substantive theory. For this reason alone Grounded Theory is important.

Knowles, 1999: 294)

However there are degrees of originality and although theoretical research can produce quite novel results, i.e. new insights into aspects of the field of study this is not the only criterion for success. One of the primary roles of theoretical research is to re-work already established ideas in order to improve insights into the subject matter. Such improvements would constitute adding something of value to the body of knowledge. This is what Proust[6] was referring to when he said: 'The real voyage of discovery consists not in seeking new landscapes, but in having new eyes.'[7] This comment is not much different to the message of Eliot (1942, cited in E.

And the end of all our exploring
Will be to arrive where we started
And know the place for the first time…

---

[6] http://www.brainyquote.com/quotes/authors/m/marcelprou129874.html viewed 8 August 2013.

[7] The danger here is that selective perception, influenced by prejudice may introduce bias.

3.7 Undertaking theoretical research

The next step is considering how theoretical research is undertaken and what steps may be made to ensure that this type of research successfully produces sound useable theories.

An eight-step approach is proposed:

1. Research question formulation;

2. Literature review;

3. Explain why a theoretical approach is being taken;

4. Concept identification and reflection;

5. Theoretical conjecture formulation;

6. Discourse with peers and other knowledgeable individuals;

7. Theoretical conjecture refinement and acceptance;

8. Discussion on the impact and implications of the theory.

3.7.1 *Research question formulation*

The starting point of all research is to establish an unambiguous re-search question. Without a clear research question, any research effort will simply wallow without direction (Winston and Fields, 2003). It is a question of *If you do not know where you want to go then any route will take you there*[8]! It is surprising just how many doctoral re-

---

[8] One of the best expressions of this is to be found in Lewis Carroll's *Alice's Adventures in Wonderland*: 'Would you tell me please, which way I ought to go from here?' 'That depends a good deal where you want to get to,' said the Cat. 'I don't much care where...,' said Alice. 'Then it doesn't matter which way you go,' said the Cat. 'So long

search projects suffer from either not having a research question or having such a poorly articulated research question that it is of little value to the research process.

The research question may be triggered by an empirical observation, although this need not always be the case, as a remark or comment in the literature can be just as good a place to start. Wolpert (1993: 6) cites Aristotle as saying:

> *'For everyone starts by being perplexed by some fact or other...'*

Research questions are seldom established after a single attempt. They evolve as the researcher explores the field of study through understanding the literature and engaging in discourse with knowledgeable informants. During this process it is valuable to articulate who will benefit from the successful conclusion of the research. These individuals will be the stakeholders of the research and they may be called upon later to comment on the research.

3.7.2 Literature review

One of the key characteristics of theoretical research is the emphasis on established ideas and concepts. Thus the researcher needs to be well read in all aspects of the literature surrounding the research question. Reviewing the literature continues throughout any doctoral

---

as I get somewhere,' Alice added as an explanation. 'Oh you're sure to do that,' said the Cat, 'If you only walk long enough.'

degree but it is especially intensive when a theoretical degree is being undertaken.

A metaphor which some researchers find useful when thinking about the literature review aspect of theoretical research is to see this research process like developing a jigsaw puzzle. This jigsaw puzzle does not come in a box with a clear picture of the required result[9]. With this jigsaw puzzle there is only a rough idea of what the final picture will look like.

Also the number of the pieces required to complete the jigsaw puzzle is unknown. Using this idea the literature review is equivalent to finding the pieces of the jigsaw puzzle. Of course the researcher will find it hard to tell when all the pieces have been found and at any one time the researcher is likely to have many redundant jigsaw pieces that cannot be ignored. Thus the researcher is truly dealing with a non-trivial puzzle.

> *Grounded Theory stood out from the rest of academic research in that it argued not to spend too much time at the outset of the research on becoming familiar with the literature. Understanding the literature can lead to an increase in the confirmatory bias. But not understanding the literature can represent ignorance. "Latter day" Grounded Theorists have softened on this issue and the importance of understanding the literature has been returned to its rightful place.*

---

[9] In academic research there is usually no unique answer or required result. We are always working within our cogitative capacity, which is limited and is subject to continuous change.

3.7.3   Explain why a theoretical approach is taken

As a theoretical approach is relatively unusual, it is important for the researcher to indicate why he or she has taken this approach to their research. The most common reason is to consolidate, extend or clarify previous works, either of the researcher or of others. In business and management studies researchers may wish to draw together different ideas from different research activities to produce a more comprehensive understanding of a situation.

3.7.4   Conception, Identification and Reflection

Once a material amount of the literature has been read, the researcher can then begin to seriously reflect on what is known about the research question and begin developing new ways of looking at the issues involved and finding new ideas to describe the relationships between these concepts. Using the jigsaw puzzle metaphor this is like beginning to fit the pieces together to form a picture. This is one of the more creative aspects of the theoretical research process and there is no cookbook recipe for this work. The researcher may obtain some help by discussing ideas, concepts and definitions with peers, colleagues and his or her supervisor(s). Reflection plays an important role. According to Alvesson and Sköldberg (2001): *'Reflection can...be defined as the interpretation of interpretation and the*

Grounded Theory is based on the type of reflection which is described here by Alvesson and Sköldberg. There is no doubt that the activity implied in being a dataist is essentially about "critical self-interpretation" of one's growing understanding. It is always a mistake to rush to understand data (It is interesting to ask if it is ever possible to rush understanding, but this may be what is sometimes referred to as forcing the data) in qualitative research and that is why the term "emergence" of theory is used.

---

launching of critical self-interpretation of one's own interpretation.'

They go on to point out the importance of reflexive interpretation and suggest that: *'Four aspects appear to be of central importance: creativity in the sense of ability to see various aspects; theoretical sophistication; theoretical breadth and variation; and the ability to reflect at the meta theoretical level'.*

3.7.5   Theoretical conjecture formulation

> The Grounded Theorist using a constant comparison process engages in the same type of theoretical conjecture as this. It has to be remembered that a theory is always only a theory and in general theories remain in place until a better theory comes along. The constant comparison process keeps the researchers' mind outward looking to see if a new and different theory is likely to come along.

The process of concept creation, reflection and theory identification will lead to the formulation of a new theoretical conjecture. Again this is a highly subjective and creative process and as DiMaggio (1995) pointed out, *'the formulation of theory is a function of our values'*. This is strongly supported by Gould (1988) when he said: *'Science is not an objective, truth-directed machine, but a quintessentially human activity, affected by passions, hopes, and cultural biases. Cultural traditions of thought strongly influence scientific theories.'*

Imagination and creativity play an important role in how the theoretical conjecture is formed as they offer access to different possibilities. One of the key tasks of the researcher is to map evidence onto potential explanations – and potential explanations are created by imagination.

Thus the more possible explanations the researcher can think of, the better. However, one must remember that imagination and creative thinking need to be tempered, as the theoretical conjecture has to be convincing and the academic community, which needs to be convinced, will be a highly critical if not actually sceptical group of individuals. Any new explanation needs to be supported by a well-argued case. Old explanations may also be eliminated by appropriate argument.

The researcher may make a theoretical conjecture at any time. The theoretical conjecture is nothing more than a researcher's suggestion as to how the ideas and relationships in the field of study actually work. Theoretical conjectures can be used again and again as stalking horses, i.e. targets to be shot down by the researcher him or herself or in debate with colleagues. This process is actually one of concept refinement and the debate engendered is an example of the dialectic in practice[10].

But finally the researcher makes his or her contribution by developing a new theory or by producing a better or fuller theory or explanation of the issues and the relationships being studied. This theoretical conjecture is the outcome of this phase of the research and will now be formally presented for scrutiny by the community and the stake-

---

[10] The dialectic, originally attributed by Plato to Socrates in his *Republic*, who called it the 'midwife of knowledge', was also used in ancient times by Aristotle. However in recent, if not quite modern, times, the dialectic was further developed by Hegel and eventually adopted by Karl Marx and others. In modern research methodology argument or disagreement usually replaces the term dialectic.

holders that the researcher has previously identified. Using the jig-saw metaphor the pieces that fit have now been put together to form a picture and the question is now how 'good' or useful a picture is it[11]?

3.7.6   Discourse with peers and other knowledgeable individuals

The new theoretical conjecture needs to be exposed to and scrutinised by other enquiring minds that are knowledgeable in this field of study. The completed theoretical conjecture will often be presented at seminars held in the Department, the Faculty or the University at large. Where applicable the researcher should also present his or her ideas to professional bodies by means of holding seminars. The researcher for a senior research degree needs to have at least one paper[12] describing his or her research accepted at a suitable academic conference. It is also useful if the researcher can have some part of the research findings published in a quality peer reviewed journal.

> *Discourse, which is sometimes referred to as the dialectic, drives research and especially Grounded Theory based research. The researcher has to be constantly asking him or herself "How convincing is my argument?" and "Are there other arguments that I should be aware of and taking into account?" Thus keep thinking and talking about the ideas that are emerging.*

---

[11] At doctoral level the creation of new theory is regarded by some academics to be too ambitious and a modification or development of an established theory is seen as being adequate.

[12] Many universities would not regard one paper as sufficient at doctoral level.

The more exposure the researcher's ideas are given the more likely different and perhaps contrary views are to emerge. It is most important for the researcher to listen carefully to these other views and to ensure that they are accommodated in the theoretical conjecture. This is again the process of the dialectic, which is essential for sound research. Collins (1994) comments on the importance of this process when he says:

> *It is important to note that there is always a judgement to be made; that scientific discoveries are not made at a single point in time and at single places and with single demonstrations. They are made through a process of argument and disagreement. They are made with the scientific community coming slowly toward a consensus.*

Sutton and Staw (1995: 373) make a similar point: 'Build strong theory over time'. Research cannot be rushed. It takes time and hard work for a researcher to derive sound results. The stories of instant flashes of research genius such as those told about Archimedes and Newton are most unlikely to be anything other than fantasies. Thus the final output of a piece of theoretical research needs to be allowed to mature in the mind of the researcher and perhaps even in the collective mind of the academic community.

On a cautionary note it is worth pointing out that research, both theoretical and empirical, will not always lead to a suitable or acceptable conclusion. Some problems are quite intractable. There are aspects of our environment that offer great challenges and sometimes it is difficult to produce a suitable theoretical explanation. Sacks (1991: 188) made this point strongly when he wrote: 'You are also going to have to bow your head, and be humble, and acknowledge that there are many things, which pass the understanding.'

With this comment Sacks was reflecting on the fact that he, even as a leading international authority in the field of psychology, was unable to understand his own personal reaction to a major leg injury he had sustained as a result of being savaged by a bull.

There are indeed limits to science and our resulting knowledge (Medewar, 1986). From the point of view of a doctoral degree it is not necessarily a disaster if the research degree candidate does not produce a new viable theory. A doctoral degree could still be obtained without a fully developed new theory in the originally envisaged form provided the degree candidate could clearly demonstrate the fact that the process through which he or she had been did actually result in something else having been added to the body of theoretical knowledge. A research degree candidate in such a position would need to make a case that the research, despite its failure to deliver a new theoretical explanation, still made a contribution. Making such a case need not be that difficult.

3.7.7   *Theoretical conjecture refinement and acceptance*
The process of discourse described above will almost certainly produce suggested amendments to the new theoretical conjecture and the researcher needs to accommodate these.

Using the jigsaw metaphor some of the pieces may not have fitted as well as originally thought and need to be discarded. Other new pieces may need to be found. This phase of the research could be short, requiring only minor amendments, or it could be extensive, needing a considerable amount of rethinking and re-evaluation of the theory.

3.7.8   Discussion on the impact and implications of the theory

Theoretical research would not be complete without a detailed discussion on the impact of the theory on both practice and on other related theories. This discussion is a major part of the research. It has been argued that the face validity of the research is reflected in the degree to which this discussion produces a convincing argument. The discussion needs to address the locus of the new theory and its impact on current thinking. Of course, in a theoretical dissertation or paper, this section will inevitably be speculative, but it could be the basis for future research in the field of study concerned.

3.8   The evaluation of research

Research is evaluated by the findings and what may be done with them. Academic research has an extra dimension which is that it not only needs the findings be acceptable to the community, but the researcher has to show a high level of scholarship. Grounded Theory demands as high a standard of scholarship as any other approach to academic research.

The evaluation of research in general is difficult. In the academic world, it is probably more challenging than in the commercial world because it is not always easy to see the immediate short-term benefits of the research findings. Of course this problem is more prevalent in some fields of study than in others. In social science research problems and questions are often directly related to solving actual problems and as such are sometimes referred to as applied research.

When the findings of this research lead to a new insight, which, it is believed, will help solve a problem, this facilitates the evaluation of the research. However in the academic environment finding a solution to a problem is not enough. In addition, the research needs to demonstrate a high level of scholarship.

In this context we define scholarship as follows:

> *The main characteristics of scholarship are that the research needs to demonstrate that the researcher has a thorough knowledge of the literature; he or she needs to clearly show that there has been a considerable amount of reflection concerning the established knowledge of the subject; and that there is a convincing argument (or rhetoric), expressed plainly and clearly in accessible language, based on a rigorous methodological process pointing to the findings. The final attribute of scholarship is that the research needs to be presented with regard for the highest standards of integrity. This means that the researcher needs to be completely honest in his or her presentation of the results.*

3.9   Evaluating theoretical research

There is not much difference between the evaluation of empirical and theoretical research. However, because of the fact that theoretical research does not rely on data or evidence collection, but on analysis and synthesis, it is sometimes said to be more difficult. With empirical research, there are better-established steps to review and techniques to assess. Theoretical research relies heavily on reflection, creativity and imagination. Although these attributes are still required for empirical research they are often required to a greater extent in theoretical research. Either research strategy can be evalu-

ated by the following tests to see if a piece of work qualifies as doctoral level research.

- Is there a clearly articulated research question, which seeks
  to establish a new theoretical understanding, refute an old
  theory or develop an extension to an old theory?
- Is the work framed within the body of current theoretical
  knowledge?
- Has the research been conducted with appropriate reflective
  procedures supported by adequate discourse?
- Has the contribution to the body of theoretical knowledge
  been expressed clearly using a convincing and reflective
  rhetoric?
- Has it been demonstrated that the new theoretical
  knowledge has some potential practical validity and utility?

3.10 Summary and conclusion

Theory underpins academe and, although there are many theories,
laws, principles, theorems and models, the idea of theory is seldom
directly discussed or explained. Because the word is so frequently
used in everyday conversation, it is often incorrectly assumed that
these concepts are well understood by faculty, researchers and examiners. There is actually a material amount of confusion about this
vocabulary, especially when it is used in the context of higher academic degrees.

A theoretical research strategy is a powerful approach to adding
something of value to the body of knowledge. It is an approach to
research which, if used correctly, can deliver material benefits to the
field of study and to the researcher. It is unfortunately not employed

extensively. As mentioned above, there are researchers, especially those who are new to the field, who believe that if there is not a set of primary data collected, then there isn't proper academic research going on. This is at best a naïve view and this chapter/paper intends to put the record straight.

At the end of the day academic research has to demonstrate that it has resulted in something of value having been added to the body of theoretical knowledge.

This needs to be done through a carefully constructed and convincing argument or rhetoric, which displays all the characteristics of scholarship discussed above. If this is done, the findings of the research will be acknowledged by the academic community as being valuable and an appropriate degree will be awarded. It needs always to be kept in mind that any evaluation of a research degree or, for that matter, a research paper needs to start and end with the view that academic research should not only be scholarly, but should also add something of value to the body of theoretical knowledge.

> *Academic research, especially at doctoral level is not for "scardy cats" and this applies especially to Grounded Theory. Grounded Theory is for mature researchers who have a significant level of experience and adequate self confidence to be able to make the case that their work has allowed them to make a contribution. It is probably good advice to say that Grounded Theory should generally speaking not be the first choice of an approach to research although it is equally sensible to say that it need not be regarded as a researcher's last choice either.*

Getting involved with a Grounded Theory project without a thorough understanding of theory and theory developments will not lead to success.

Some Grounded Theorists suggest that because Grounded Theory most often only involves substantive theory that the detailed knowledge of theory development is not really necessary but this is simply wrong. Even at the substantive level Grounded Theory requires a sophisticated knowledge of theory development.

Of course it requires a lot more study than can be delivered in one chapter or paper to become knowledgeable about theory and theory development from an academic perspective.

4

Data?

Data is spoken of by nearly every researcher but understood by few.

Glaser famously remarked "All is data" and thus the importance of data to Grounded Theory is hardly surprising. Of course this comment without extensive explanation is not necessarily helpful, especially to novice researchers.

It is really surprising how little has been written on what constitutes data in the context of academic research. It is of course addressed by statisticians and also by some quantitative researchers. But in general it is a blind spot and a number of academics have difficulties in discussing this concept. Some academics fall back on the hierarchical model created by Ackoff (1989) which postulated that there was data, information, knowledge and wisdom, but even if this is accepted it is not an adequate explanation of the nature of data and how to distinguish it from that which is not data. Ackoff's contribution may have been useful to computer scientists, but was not of much help to academic researchers.

As an aside it is also interesting to reflect on how frequently a discussion of data raises the comment that data is a plural noun and therefore it is not correct to say or write "data is". Latinists are always pleased with this rather irrelevant and increasingly rejected view of the word data.

When talking to academics about the nature and characteristics of data it is often assumed that it is clear to everyone what data is, but when some questions are asked about it the replies obtained can be fussy if not actually confused. Few researchers focus on that one essential attribute of data which is that data is that which facilitates the answering of the research question.

Grounded Theory is a dataist approach to research and to understand what dataism is about it is important to have a thorough grasp on the nature of data from an academic research point of view. This is not a trivial matter as there are many different dimensions to data and there is not universal agreement on them and thus this chapter requires careful reading and reflection.

The paper which constitutes this Chapter addresses the definition of data and discusses data measurement and data integrity together with a number of other important issues. Like Chapter Three the discussion provided here goes beyond that which is needed for the purposes of Grounded Theory.

This paper was originally written as an occasional paper by Dan Remenyi.

Data?

Data is clearly an ideal life form. Who cares if it (or maybe he should it be?) is synthetic? Immune to nearly all biological diseases and with the strength of many humans Data has saved the good star ship Enterprise as an officer of distinction on many voyages. He is thought of fondly by his fellow bridge officers. (Anonymous)

4.1 Introduction

An interesting gap in the research methodology discourse has been created by the lack of a convincing and useable definition of data. It seems that data is such an intensely used concept that it is often assumed that any and all researchers will be able to identify data and know the issues related to its use and the conditions under which data may be considered valid.

This is surprising considering how important data is to research. Most research projects undertaken today are empirical and thus rely heavily on data. Even when the research is not empirical i.e. theoretical research based on argument, some of the ideas and theories which are the raw material input to this research process will have been arrived at through some consideration of data. An interesting side effect of this lack of attention to the nature of data is frequent misunderstandings about this subject even among the research and philosophical elite and the comment made by Russell (1960), *"data can only be criticised by other data"*, illustrates this. This statement suggests a high degree of confusion about data. Data can be shown to be irrelevant and or invalid in a number of different ways and such arguments need not rely on other data.

The assumptions or the lack thereof made by most researchers about data have sometimes resulted in a poor understanding of the issues involved with the research question. Rushing to collect data, without adequate reflection on what data is required to answer a research question is indicative of a technician approach to research (Gummesson 2000) rather than the scientific understanding which is actually required.

In the field of business and management studies and some other aspects of social science, researchers, when asked to define data, will sometimes rely on the hierarchy developed by Ackoff (1989) which is that there is data, information, knowledge and wisdom and this is sometimes referred to as the DIKW pyramid model or hierarchy. This model portrays a progression from data to information to knowledge to eventually wisdom without offering an adequate understanding of data. At best this model is superficial and at worst it is misleading. It is of little value in defining data from an academic research point of view as it neither provides a comprehensive understanding of the nature of data nor does it help with an explanation of the issues related to the identification and collection of data.

The objective of this chapter/paper is to outline the landscape in which data exists and is used and to give some guidance as to how to think about and use data effectively in answering a research question.

4.2 Data and measurement

In research, data is often regarded as the result of a measurement. Sometimes the measurement occurs during an experiment and sometimes as a result of observation. Measurement involves assigning a number to some aspect of a phenomenon in which we have an interest and this number represents its relative position on a scale which is in use (Cooksey 2012). In the physical and life sciences laboratory experiments may be designed over which the researcher has extensive control. In these circumstances the measurement of the variables concerned can be relatively straightforward and the issues related to the data so acquired relatively unproblematic. Data which has been produced as a result of experimental measurement is often described as hard data and it has become customary to regard hard data as more accurate and more reliable and amenable to statistical analysis and mathematical modelling. The importance of measurement was articulated by Kelvin (1900) where he said:

When you can measure what you are speaking about, and express it in numbers, you know something about it; but when you cannot measure it, when you cannot express it in numbers, your knowledge is of a meagre and unsatisfactory kind.

Hardly anyone would argue that Kelvin was wrong especially for his time. But more is required of research in the 21st century and we are also today more wary of the potential problems always associated with measurement.

This attitude is still largely prevalent among researchers although there is increasing awareness that there is much more to data than only numbers.

Data obtained by observation in the physical and life sciences may be the results of measurement, but it may also be descriptive and thus not amenable to measurement. The biological and geological sciences are examples where descriptions constitute a major form of data.

Some researchers argue that the distinction between hard and soft data is fundamentally misleading. This has been exacerbated by the fact that many people think of numbers being hard and descriptions and other forms of data which are not numbers as being soft. Maybe it might be better to say that data in which we have a high degree of confidence is hard and that data in which we haven't is soft. This might make more sense if we say that findings based on data of which we are confident produces harder or more convincing results. However this would require us having to admit that we sometime use data that might be regarded as not having provenance.

In the social sciences measurement can play a lesser role. In researching human and organisational behaviour only a relatively small amount of the data available may be of the kind which is amenable to what is normally regarded as measurement. Data based on description tends to be more abundant. This is not to say that there is no measurable data available such as that which may be acquired from questionnaires or from organisational records, including financial accounts, production and sales and human resource reports to mention only a few sources. But this type of data should not be confused with that which is usually obtainable from experiments and described above as hard data.

Numeric data in the social sciences, which is sometimes referred to as hard data, may not be accurate and as reliable as it appears to be. An example of this is the way that ordinal data based on opinion surveys is sometimes treated as though the data represents measurement of physical objects. Irrespective of the nature of the research being undertaken, the veracity and integrity of data should not ever be taken for granted and it should always be questioned.

There are a couple of other important points to be made about measurement. Firstly what is measured is decided by the researcher. What one researcher believes to be important another may not and thus this decision may directly affect the outcome of the research. There is then the issue of the measuring instrument and the unit of the measurement. These are only a few of the issues which can make a substantial difference to the outcome of research based on measurement.

4.3 The research contextualisation issue

As a result of experimentation in the physical and life sciences, especially when statistical or mathematical modelling is concerned, the data employed by researchers will be predominately numeric. But this is not the only form that relevant data takes and which researchers will often have to deal with. Researchers have to contextualise the statistics or the mathematical models they produce. Numbers alone are too parsimonious and despite the aphorism that facts speak for themselves, this is often untrue. Research findings have to be interpreted and other types of data are required for this. This point was made clear by Paulos (1998) when he commented:

> But even where mathematics or statistics is the preferred way of understanding the situation, 'without an ambient story, background knowledge, and some indication of the providence of the statistics, it is impossible to evaluate

93

their validity. Common sense and informal logic are as essential to the task as an understanding of the formal statistical notions.'

To be able to bring meaning to the results or findings of mathematical or statistical analysis the researcher needs to be competent in understanding and explaining the assumptions made, the context in which the research has been conducted and the ways in which the findings may be employed. This requires the ability to comprehend the nature of the environment by accessing other data found therein and to make sense of both numeric data and non-numeric data.

As well as the physical environment in which the research was conducted it may be important to be aware of the attitude of people who took part in the research and this will usually require non-numeric data. Data collected concerning the context of the research is sometimes referred to as soft data. Soft data is generally considered to be less reliable than hard data as soft data may include opinions and descriptions that will inevitably be subject to bias.

It is certainly possible to find contextual data which is quite hard. For example when researching on an industry level a researcher may wish to use GDP figures as well as perhaps CPI or the RPI figures. Many researchers would regard these types of data as being hard.

There is a wider range of data available to the researcher than is frequently credited. In most academic enquiry a researcher will use both hard and soft data. The decision as to which type or types of data to acquire is largely determined by the research question and also on the researcher's ability to access appropriate data. These issues will be discussed in some detail later in this paper.

4.4   A definition of data

In exploring the issues related to data it is useful to have a comprehensive definition of data. As a social scientist there is the temptation
to list data types such as numbers, words, sentences, letters, conversations, i.e. verbal and written communications, and offer this as a
definition. However with a little reflection these examples may be
generalised by stating that

> *data may be considered to be any sense perception that*
> *the researcher receives and which he or she believes will*
> *be helpful in obtaining a fuller understanding of, or an-*
> *swer to the research question.*

Thus a sound could result in data or a smell or the feel of an artefact
in the hand or an observation of a behaviour or artefact and so forth.
Within the Grounded Theory tradition Barney Glaser is well known
for his dictum *"All is data"* (Bryant 2009) and although this is correct
it is not especially helpful without further explanation. What is important is that data results from the sense perception and it is not
the event itself. Data may be perceived and remain in the mind of the
perceiver or it may be perceived and recorded on some medium for
further consideration including analysis, interpretation and reflection. Data becomes most useful when it is recorded, if for no other
reason than it gives the researcher a greater opportunity of recalling
and discussing the event. Some researchers argue that until it is recorded an experience has not become data. However it would be
better to argue that it is difficult to access the usefulness of data until
it has been recorded.

This first definition provided above, which should not be regarded as comprehensive, draws attention to the wide range of stimuli that can constitute or be considered data and which at the same time alerts us to the possibility that not all stimuli will be or could be regarded as potential data. In the context of academic research the issue of the relevance to answering the research question is of primary importance. Stimuli which the researcher receives, but which in his or her view do not contribute to being helpful in obtaining a fuller understanding of, or answer to the research question may be regarded as irrelevant or *noise*. The notion of noise is borrowed from electrical and sound engineering where it refers to unwanted sounds that are of no value to the listener and which will in general interrupt the listener's ability to receive the sound he or she wished to hear.

In academic research data can only be understood in relation to how it can contribute to answering the research question. And the problem is that when the researcher is draining the swamp and fighting of alligators the primacy of the research question can be forgotten. So in a broad sense noise could be considered to be a high level issue which involves anything that distracts the researcher of which only one aspect is inappropriate "facts or figures" etc.

In research the notion of noise is used in a similar way to the description above except that such noise can be received by all five senses and draws the researcher's attention away and in so doing obscures the important data required. There are different ways of thinking about noise. In the first place noise may be the result of simple misunderstandings between researchers and their informants. Noise may also appear as too much apparent data resulting in problems related to deciding which data should be used. But there is also another use of the word noise in research data and that occurs under the following circumstance. If a Likert type scale is being used it is well established that one person's rating of say 5 out of 7 might actually represent the same attitude as another person's rating of 6 out of 7. This type of difference is sometimes described as noise. When this type of situation arises there is usually a sizeable sample, which it is believed will help resolve such noise.

Furthermore, in academic research an issue similar to noise is the integrity of the data and the researcher needs to be vigilant that he or she minimises the possibility of their being mislead, intentionally or accidently, by that which appears to be data, but is not.

As a consequence of this it is necessary to admit that whether something is considered to be data is in the hands of, or perhaps rather in the mind of the researcher and what one researcher may consider being appropriate data another might believe to be irrelevant. With regards this issue there is no uniquely qualified arbitrator to whom to appeal. There will however in due course be the voice of the research community when the results of the research are published or offered for examination and this is when matters such as the appropriateness of the data are normally judged.

From an operational point of view researchers have to answer for themselves the question, *What criteria should a researcher employ in deciding whether a particular sense experience provides data – and data that are useful/relevant?* There is no simple answer to this difficult question. Successful researchers, i.e. those who have been published in highly rated journals, will answer that the knowledge required to differentiate relevant data comes with experience. Al though this is almost certainly the case, it does not really help in identifying useful principles for the decision processes involved. Researchers have to be particularly conscious of the possibility of acquiring incorrect or irrelevant data.

There is one further dimension which needs to be addressed at this point and that is the fact that in extreme cases researchers will not be able to perceive relevant data due to their current belief or mind-set. Researchers are always required to be open-minded and thus accept new ideas and engage with new concepts which arise from the research. Not everyone will be adequately open-minded.

No matter how closed-minded a person is few would consider themselves to have such an attitude and even less would admit to it. Is it possible to be too open-minded? Some researchers argue that a refusal to easily change one's mind is a sign of confidence in one's beliefs and this may be seen as a form of integrity and as such is regarded positively.

Being open minded is always considered to be a "good" thing for researchers. But can one be too open minded? The answer to this is Yes! Changing one's mind too readily is sometimes a sign that the issues were not thought through carefully enough. Also sticking to one's point of view is sometimes regarded as a sign of integrity. Of course this is not always the case. The issue is perhaps that researchers need to be prepared to listen and look at new evidence carefully and if it suggests that a new paradigm is required they should give this careful consideration. They need to have engaged in enough reflection to be able to craft an argument in support of the decision they make after considering the evidence.

There is also the question of imagination. Researchers need to be able to imagine possible outcomes of their research in order to be

able to recognise data which might otherwise be missed by the research process. Ray (1993) was referring to this point when he said
that:

> *We are beginning to realize that if we don't believe in*
> *something, it doesn't exist - no matter how much data is*
> *thrown in front of us.*

This is a rather bold statement, and for academic research needs rewording. But the point is clear and is often valid. If we are not aware
of the possibility of a phenomenon then how could we even start
looking for it or recognise it when we encounter it? Having copious
data alone does not lead to research success especially with regard
to developing new ideas.

4.5 Data and inference

The above discussion is only a starting point for a comprehensive understanding of data. Oreskes et al (1994) cited in Horgan (1996)
moved the subject forward when they said:

> *"What we call data," they explained, "are inference-laden*
> *signifiers of natural phenomena to which we have incom-*
> *plete access. Many inferences and assumptions can be*
> *justified on the basis of experience (and some uncertainty*
> *can be estimated), but the degree to which our assump-*
> *tions hold in any new study can never be estimated a pri-*
> *ori. The embedded assumptions thus render the system*
> *open." In other words, our models are always idealisa-*
> *tions, approximations, guesses.*

This extends the previous definition by introducing a number of important concepts. The main issue introduced here is the notion of *'signifier'* on which some elaboration is required. Everyday life is full of signifiers in terms of signs. There are street signs, road signs, safety signs etc, and these are probably the most significant means available to a traveller when considering how it is possible to find a route from one place to another. A full discussion of this requires an investigation of semiotics which is beyond the scope of this paper, but it is useful to bear in mind the inherent sign nature of data. Academic research is often referred to as a voyage of discovery, which of course is a voyage of the mind, and as such the researcher needs intellectual signifiers or signs with which to find an appropriate way. Born (1950) using the metaphor of the journey remarked:

> *There is no philosophical high-road in science with epistemological sign-posts. No, we are in a jungle and find our way by trial and error, building our road behind us as we proceed. We do not find sign-posts at cross-roads, but our own scouts erect them to help the rest.*

Asking questions and obtaining answers is after all one of the primary ways non-experimental data is produced and is therefore central to this process of sign-post building or finding the way. Data, or rather the researchers ability to associate meaning to data, triggers ideas as well as offers some degree of assurance that the researcher is on a useful path. If this journey is undertaken without data then should the destination be reached, it will be only by chance. Of course not all researchers would completely agree with this opinion, which could be referred to as a dataist view (Alvesson and Sköldberg 2008). The term dataism is sometimes used to suggest that too great an emphasis has been placed on data when coming to a research finding. This is useful in reminding researchers that it is the ability to understand and to associate meaning to the data which is the central issue in research.

The second important point raised by this definition is the fact that data can be *inference-laden* and that some estimation of the uncertainty associated with potential inferences can be estimated. This definition also exposes the fact that any data set, and this is of course especially true for academic research will almost inevitably be incomplete. This issue is closely coupled with certainty and uncertainty. In academic research, certainty is always illusive and thus the question becomes what level of uncertainty or confidence in the results is acceptable. The level of confidence will be a function of the field of study and the techniques performed in order to acquire the data. In business and management studies, when statistical testing is involved, researchers often use a confidence limit of 95%.

The notion that too much emphasis can be placed on the data in coming to a research conclusion is an interesting one. It is generally accepted that both Galileo and Mendel were somewhat economic with the truth when it came to understanding and interpreting their data. Nonetheless it is also accepted, without question, that their findings were correct. The data which Einstein had at his disposal for his work on Relativity was not much more than the outcomes of his thought experiments. What might it be like to travel on a beam of light? He asked. Of course he was not trying to be an empiricist, never mind a Grounded Theorist!

Of course it is not appropriate to think that all data could be used inferentially, at least in the formal sense of that word. For inferential statistical models to be valid it is necessary that the research conforms to a number of important assumptions (Remenyi et al. 2011) and it can be difficult to establish that these assumptions are being adhered to and the assumptions need to be in the mind of the researcher as the research design is being developed. But even from the point of view of statistics much can be achieved through the use of descriptive statistics which requires little computation.

4.6   Sources of data

In business and management studies the sources of data that most researchers will encounter will most frequently consist of a number of the following data types.

- Observations – visual, audio and otherwise;
- Interviews – one-on-one and sometimes in groups;
- Conversations with individuals or in a group;
- Focus groups;
- Corporate reports, financial and others;
- Completed questionnaires;
- Websites including social software systems;
- Blogs;
- Data Caches including;
    - Collections of letters, memos, e-mails or other correspondence;
    - Articles in newspapers and magazines;
    - Documentaries – film and video;
    - Still photographs;
    - Fictional literature;
    - Speeches published or broadcast on radio or TV;
    - Diaries;
    - Other ITC databases.

Some researchers would claim that observation is the best form of data as the researcher is seeing and experiencing for him or herself what is happening, but unfortunately this is not always possible. The event being studied may have already happened or the organisation may not allow the researcher physical access to the business and management processes. Consequently the data may have to be collected through interviews, focus groups, questionnaires or some other research activities. These activities are normally based on individual informant's recall of events and as such they can be unreliable or at least less reliable than direct observation. But even with direct observation it is possible that the researcher may not perceive the events in a way that would be regarded by other observers as being objective.

A question which is sometimes asked is, *Can a research degree be obtained by acquiring only one type of data such as interviews?* Although there may well be cases where this has happened, good research practice requires data to be confirmed by triangulation wherever possible and thus multiple sources of data are quite strongly preferred. Spoken and written data complement each other well and of course there are different types of spoken data and written data as well.

It is always difficult to know in advance where the search for data will take the researcher and thus all opportunities to collect data should be taken, although researchers need to be aware that it is possible to be overwhelmed and care has to be taken to ensure that whatever is obtained can be of direct use in answering or acquiring a better understanding of the research question. Some researchers have difficulty in bringing to an end the data acquisition process as they begin to believe that the next interview or the next focus group will really reveal the answer to the research question. When enough data is acquired this process needs to be halted. Of course it is a judgement call as to when this point has been reached.

4.7 Data integrity

The integrity of any data set is always of central concern and it needs to be addressed if the research is to be recognised as having value. The data could be said to have integrity if it has been acquired from an appropriate source with suitable checks that it is both valid and relevant. As the custodian of the integrity of the data the researcher has to carefully monitor on a continuous basis the credibility of all the data obtained. Where possible the researcher should seek corroborative data especially on important issues. If this cannot be done then the researcher needs to comment and explain the possible reasons for this. And of course there are always the cases of deliberate fraud. Data has been changed or even invented from Claudius Ptolemy dating around 150 CE onwards. Broad Wade (1985) supplies a number of interesting anecdotes but as Gould (1992) pointed out

Fraud (in science) is not historically interesting except as gossip

It is important that the data has been correctly and accurately recorded and stored on a medium which is easily accessible and which will not deteriorate in quality. The record of the data should be complete and not have inconvenient elements omitted. The data should have been collected over a reasonable time span and not be so old that any of its relevance has been lost. The researcher should not include any data which he or she believes not to be correct or that contains exaggerations or a bias which might adversely affect the outcome of the research. Errors in data management, processing and retrieval can also become an integrity issue. Many different types of error can be introduced during the research process and it behoves the researcher to be on the lookout for these and to ensure that the data is handled in such a way that there are few if any errors.

This is such an important issue that if any elements of the data integrity are called into question then the whole research project will be in danger of being declared flawed and hence null and void.

The confirmatory bias is a well known phenomenon whereby researchers can unwittingly be searching and acquiring data to support their beliefs and this non-conscious bias directly affects the results of their research. In academic research bias can be difficult to identify, monitor and compensate for as researchers will not be conscious of the attitude which drives the bias. Introducing a regular set of reflective breaks in the collection of the data can be helpful to surface this issue. At the same time the researcher needs to regularly discuss his or her research ideas with others at events such colloquia where such biases may become apparent.

The confirmatory bias can cause serious threats and challenges to the integrity of the data and thus to the whole research project.

4.8   More on hard and soft data

In the social sciences the distinction between hard and soft data can be difficult to justify. Hard data will often have hidden or at least non-transparent assumptions behind the numbers. When counts are involved such as the answer to *How many?* there can be a question of whether certain units or instances should be included in the count. An example of this is *How many outstanding sales people does the organisation employ?* which requires a definition of *outstanding* and it may be difficult to establish consensus in this respect. In this question the word employ may also be problematical in an environment in which outsourcing has become so popular. Even more straightforward questions have to be answered with care. Financial figures are based on assumptions and accountants will readily admit that financial figures are produced in terms of the Generally Accepted Accounting Principles, which have considerable scope for flexible implementation and thus two different accountants could produce two different sets of figures to represent the same financial situation. The degree of uncertainty can sometimes be managed, but it cannot be totally eliminated.

There is sometimes a false illusion of certainty with numeric data. The term *hard data*, meaning *certain* data is used inappropriately as a result of this. One of the most important features of the work of the researcher is to become aware of the uncertainties involved, especially with the data and to evaluate to what extent they impact the research and then to report on this. The research should be creating field notes concerning the veracity and the validity of the data he or she is collecting. These notes could be regarded as metadata.

The term soft data, also referred to as qualitative data, is used to describe any data which is not presented as numbers. Qualitative data is used to drive qualitative research, which is a term that is not always well defined. Strauss and Corbin's (1998) definition, although succinct, is sometimes thought to be too parsimonious:

By the term 'qualitative research' we mean any type of re-search that produces findings not arrived at by statistical procedures or other means of quantification.

Defining a concept in terms of it not being anything else is unsatisfactory and the superficiality of the above definition demands a much more insightful explanation of what is meant by qualitative research, which is a non-trivial approach to research.

Denzin and Lincoln (2003) provide an overview of the components and practices of qualitative research:

Qualitative research is a situated activity that locates the observer in the world. It consists of a set of interpretive, material practices that makes the world visible. These practices turn the world into a series of representations including field notes, interviews, conversations, photographs, recordings and memos to the self. At this level, qualitative research involves an interpretive naturalist approach to the world. This means that qualitative researchers study things in their natural settings, attempting to make sense of, or to interpret, phenomena in terms of the meaning people bring to them.

It is important to recall the comment of Paulos (1998) who said that mathematical and statistical analysis alone are not enough and that researchers need to contextualise their research. It is this approach of looking for a broader understanding which is referred to as qualitative research and this allows contextualisation to be taken into account.

Thus descriptions, conversations, photographs or videos are all part of the rich data available. But in what respect should these types of data be considered soft? Some researchers mistakenly feel that the word soft is somehow a synonym for easy or trivial. This is not the case, as qualitative data can be difficult to acquire, record, analyse and understand. The use of the words *soft data* mostly describes the attitude of the researcher towards the data and not any aspect of the reality of research, and thus perhaps the terms hard and soft data have outgrown their usefulness? There is still a distinctly negative attitude towards qualitative research in certain quarters. This is based on the mistaken belief (in modern times going back to at least Kelvin, as mentioned earlier) that if something cannot be measured in such a way that it can be treated by statistical analysis or mathematical model it is of lesser value. Kennedy (1979) illuminated the falsity of this misconception when she said:

> *It is important to realise that non-statistical arguments need not be invalid. Yet many researchers may be timid about attempting such inferences simply because the rules as to what constitutes reasonably sound inferences are ambiguous, relative to the rules as to what constitutes a sound statistical inference. What is needed are rules of inference that reasonable people can agree on.*

The problem faced by researchers is that the rules relating to inferences from qualitative data are the rules of logic and argument. These are generally not taught in schools or universities and are thus not well known or understood. These rules are also amenable to subjective manipulation.

4.9   Data collection, gathering or creation

Data is commonly regarded as either primary or secondary. Primary data relies directly on the sense perception of the researcher, mostly in the field and has not existed before. It is the research process that allows the data to be obtained and to be used in the processes required to answer the research question. The act of collecting the data will normally reflect in some way the assumptions of the researcher. Secondary data is that which has already existed and has been produced by means other than the efforts of the researcher and which is normally already published and consequently accessed in written form. The production of this data will not be influenced by the assumptions of the researcher, but the decision to seek and find such data will be so influenced.

There is an important distinction between naturally occurring data and elicited data. The term naturally occurring data is used to describe data which existed before the research began and all the researcher had to do was identify it and access it. Naturally occurring data is in some ways similar to secondary data, although as mentioned above it is usual to perceive secondary data to be in some formal sense previously published.

An example of naturally occurring data would be data held by cash registers in supermarkets for the purposes of customer checkout and accounting. This data would not have been recorded for research purposes and it would not have been published.

It is common practice to refer to naturally occurring data being collected or gathered by the researcher, but this may not be the case for elicited data. Elicited data is sometimes said to be created or generated or even produced. In business and management studies data is elicited by the researcher asking a specific question of an appropriate informant and the answer or data acquired to this type of question may be highly influenced by the way the question is put and to whom the question is addressed. This may be done face-to-face or by some remote technique such as a self completion questionnaire. For this reason interview schedules and questionnaire design and the selection of an appropriate sample frame are so important. Self completion questionnaires are always subject to a self selection bias.

When data is elicited from informants the researcher has to be aware of the possibility of the question not being fully understood. It is also possible that the reply may not be adequately articulated and therefore possibly misleading. Human reactions around the question and answer process become a central issue in collecting qualitative data and therefore the meaning of communications between the parties concerned cannot be taken for granted. The researcher needs to be alert to this issue and be ready to explore issues when it is necessary to challenge the apparent meanings. Triangulation can play an important role in verifying the fact that the different parties have understood one another. But it is also important to appreciate that even when data is confirmed by triangulation it can still be wrong. For this reason the researcher needs to continually challenge whether the data is valid, appropriate and useful in answering the research question.

The researcher needs to be aware that data should be timely. Data collected years earlier may not have adequate relevance except in a historiography study. Universities are now much stricter with regards to completion times. Some researchers have taken breaks of multiple years from their research and are surprised to hear that their data may no longer appear relevant. The exception to this would be a longitudinal study and the extended length of time would need to be agreed at the research proposal stage.

4.10 Academe and research questions

The research process revolves around the research question and the research question drives the data required. It is not always a simple matter to determine the data requirement at the outset of the research program and frequently data issues have to be revisited at different points of the research process – research programs seldom follow the waterfall model of project management. The researcher will be learning throughout the duration of the research and will be acquiring more skill at knowing what data is required. Not only can data be associated with individuals and organisations or departments thereof but it may also be associated with artefacts and events, for example a new computer system and its acquisition and implementation. In such instances it becomes ever more important to be clear on which types of data are required and how they will be obtained. It is sometimes difficult to know where suitable data will be found. It may need to be detailed or focused and at the same time general enough to be able to perceive the big picture. It needs to be quantitative and qualitative. In order to employ a mixed method approach (Clark and Creswell 2008) it may need to represents facts and opinions. It may need to come from multiple sources.

One of the important issues in deciding the data requirements is to have a clear view of the unit of analysis that the researcher is examining. Although this is a central issue in any research project there is sometimes confusion about this aspect of the research. If the researcher is seeking to answer a question relating to an individual role or function, then it is the role or function that is the unit of analysis and detailed information about the activities performed in that role or function will be required. If the question refers to the functioning of a section or department then perhaps information about the objectives, performance and evaluation of the section or department is the central issue. Sometimes the unit of analysis will be at corporate level. It is critical that the issue of unit of analysis is clarified as early as possible in the research as it is a useful indication of the data sources required. In some cases there will be multiple units of analysis and so a number of different data collection strategies may be required.

4.11 How much data is enough?
Each data analysis strategy will have specific data requirements and researchers need to be aware of these. If an inadequate amount of data is acquired then a particular data analysis may not produce acceptable or even unusable results.

There a general belief among some researchers that it is not possible to have too much data. However this does not take into account the challenges and the costs associated with data collection. Collecting more data than is necessary will delay the finalisation of the research. Having more data than is necessary opens the researcher to a greater possibility of error in both recording and processing, and it increases the cost of the research. The term data asphyxiation has been coined to describe the case of extreme data overload which can distract a researcher from the objectives of the research.

The adequacy of a data set is a considerable worry for many researchers. In general research it is often decided in advance how may data points will be obtained such as, You will need to have 100 completed questionnaires, if your research is to be credible. In Grounded Theory the need for data evolves as the researcher proceeds. Thus a statement such as the one just made would be an indication of the researcher not understanding the basic principles of Grounded Theory. The need for data continues to evolve until the researcher has a comprehensive understanding of the emergent categories and constructs which are required for the theory.

But on the other hand, perhaps paradoxically there it is a fact that the researcher may have to be satisfied with incomplete data. The expression of *What You See Is All There Is* is hardly ever correct and competent researchers are able to detect when they are missing potentially important data.

Whereas quantitative data analysis techniques have clear rules regarding the size of the data set required the issue is not as clear when it come to qualitative data analysis. Quantitative data analysis often relies on obtaining a sample of data from a greater population. This approach is only relevant if it is the intention of the researcher to make inferences from the sample to the population as a whole from which it was drawn. Nonetheless, qualitative researchers who generally do not make such inferences will sometimes use this term even when referring to the number of interviews they have acquired. Strictly speaking this may not be the correct use of the word sample.

Qualitative researchers employ the concept of data saturation to determine when enough data has been acquired and this necessitates a judgement on the part of the researcher. This is not a trivial issue and researchers need to be able to argue convincingly that data saturation has been achieved. An account of the issues related to data saturation is provided by Grant (2012).

Some researchers find data acquisition almost compulsive and spend too much time on this aspect of their research. When this occurs it suggests a lack of understanding of the requirements of the research process and a level of insecurity in the researcher. An important rule of research is that no amount of data can substitute for insightful understanding and interpretation of the variables relating to the research question. A small data set in the hands of an insightful and imaginative researcher can produce quite satisfactory results.

4.12 Data and insights

Within some business and management research studies data collection will constitute a major part of the total research work, whilst in others it will be a relatively quick and easy process. As each research project is so different it is not possible to prescribe a guideline in this respect, but researchers can count on the issues related to the data being of central importance and consuming an important part of the time allocated to the research. Scott Fitzgerald (1920) commenting on how long it took him to complete a book once remarked:

> *To write it, it took three months; to conceive it—three minutes; to collect the data in it—all my life.*

In a sense this describes academic research well, as an academic research project is inevitably built on the accumulated knowledge and understanding acquired by the researcher up to that point in his or her entire working experience. The researcher's thoughts, influenced by his or her pre-knowledge which is in at least some senses a form of 'data' will have been developing in his or her mind for some time. The researcher exposes his or her sensibilities to the question and the environment under which the study takes place and collects whatever data he or she considers appropriate. The data so recorded needs to be as accurate as possible. When the data is understood, interpreted and used to create a theoretical understanding, it needs to be done with the voice of the researcher. It is important that a supervisor or a reviewer does not intervene too much in the voice of the researcher as this will cause the research to lose authenticity.

Some researchers report flashes of insight when they encounter some aspect of the data and a question sometimes posed is *"Is the data that produced a flash of insight in the researcher different in any way to more routine data which is collected, managed, analysed and interpreted and which then may produce some new way of thinking about the research question?"*

In general researchers would not see any material differences in the type of data which could be involved in the insight. It has been suggested that insights are the result of some internal cognitive dialectic process which the researcher embarks upon in his or her own mind and is not the result of some special sort of data obtained. Stories of insights abound in science. Perhaps the best know example is that of Newton and the ap-

When it comes to qualitative research it is understood that there is no clear line to be drawn between data collection and data analysis. The researcher is involved with data understanding and interpretation while he or she is collecting the data. There is no intrinsic problem here as the data collection process is a useful learning undertaking as long as it is understood as such. However the idea of collecting data first and then analysing it and then moving on to interpreting the data is not a useful model in qualitative research.

ple falling on his head after which he conceptualised and understood the laws of gravity. But this is seldom regarded as authentic. Einstein's story (Ishiwara 1977) on the other hand told by himself about himself seems more authentic, "I was sitting in a chair in the patent office at Bern when all of a sudden a thought occurred to me: 'If a person falls freely he will not feel his own weight'. I was startled. This simple thought made a deep impression on me. It impelled me toward a theory of gravitation".

4.13 No data is sometimes data

An interesting aspect of data is that the absence of a sensory stimulus can constitute useful data. By far the most well know example of this in English literature is the exchange described in *Silver Blaze* which is one of the stories in the *Memoirs of Sherlock Holmes* (1993). In this story there is an exchange between Holmes and a Scotland Yard detective which goes as follows:

> Gregory (Scotland Yard detective): "*Is there any other point to which you would wish to draw my attention?*"
> Holmes: "*To the curious incident of the dog in the night-time.*"
> Gregory: "*The dog did nothing in the night-time.*"
> Holmes: "*That was the curious incident.*"

This type of recognition of the meaning of silence is prevalent in our society and is important to researchers. If an apology is made and the person to whom the apology is addressed does not reply it can generally be understood that the apology was not accepted. If an organisation is accused of transgressing regulations or being unconcerned about safety for example, then no response to this could be regarded as an admission that they have been in the wrong, but cannot bring themselves to apologise.

If a researcher asks an informant a question and the informant ignores the question then the researcher is entitled to consider the lack of response as perfectly sound data.

4.14 Matching research questions and data

Research projects can seek a wide range of data which could be useful in answering the research question. Typically the potential range of data will be wider than is practical to fully address and there is likely to be more potential data available than the researcher will have resources to obtain. One of the skills which has to be acquired by successful researchers is to be able to discern which are the most important and useful sources of data and then to obtain the data with as little effort and cost as possible. In this respect the 80/20 rule will frequently apply. The researcher has to be cognoscente of the need for efficiency. It is difficult to generalise about this as every case will be quite different.

The nature of the data required for a research project is normally considered to be a function of the research question. Some research questions lend themselves to quantitative data analysis while others need more detailed data than can be more appropriately obtained by acquiring measurements and thus more qualitative data is necessary. However as well as the research question, the background and skill of the researcher is an important issue in determining the data required. Some researchers who have a quantitative background such as engineering or science may prefer to work with quantitative data while others with, for example a humanities education may feel that qualitative data is the way forward. This professional bias is critical in the data type selection decision.

4.15 The question of access

But perhaps most of all the data collection decision is always related to issues of access. No matter how insightful or helpful a data set might potentially be in answering a research question, if the researcher cannot obtain access to it then it is not worth spending any time considering it. Most researchers find data access to be challenging and sometimes they are surprised at how unhelpful potential informants can be. Gatekeepers (Remenyi 2012) can be helpful, but sometimes it is difficult to find such individuals.

When considering access there are two dimensions to take into account. Besides the obvious issue mentioned about of how to approach an organisation and the benefits of finding a gatekeeper there is the subsequent problem of how to convince individuals in that organisation that they should be open and discuss frankly the issues which concern the research. This is not a trivial matter as many organisations do not have an open culture which will not tolerate an employee speaking freely if there is any direct or even indirect criticism of its management. In many organisations there can be severe consequences if management becomes aware that an employee is not 'loyal'. In this sense all employees are vulnerable. Certainly, the researcher will as a matter of course provide assurances of confidentiality, but will this be enough to allay any fears the informant might have of negative repercussions? Recently a researcher has produced an interesting way of describing what can happen with regards to informants not being prepared to speak their minds openly. She reported that she was conscious that the informants she had interviewed were only prepared to repeat the 'corporate cassette' i.e. the formal company line. Such 'data' if it is correct to call it data, is of little value for most research. In instances such as this the researcher may asked the informant to be more open and if this does not produce a changed attitude the researcher should not pursue data collection from such an individual further.

Once again there is no simple answer to the conundrum of access and the openness of individuals. The researcher's ability to provide creditable assurances of confidentially to informants is central and thus this issue may end up as a function of the personality and the maturity of the researcher.

4.16 A checklist for a data acquisition plan

During the development of the research proposal and/or the research protocol time needs to be devoted to considering the issues mentioned in this paper and to producing a plan of action which will ensure that the data collection aspects of the research are managed as efficiently and effectively as possible.

The following is a list of questions that researchers may find useful in scoping the data acquisition dimension of the research and which should be incorporated into either the research proposal or the research protocol or both. This exercise can only be commenced when a clear research question has been established. Research questions often require their reduction into a number of sub-questions and this also needs to be completed before the following issues can be competently addressed.

1. *Which data sets are likely to facilitate the answering of the research question?*

2. *How many different data sets may be involved and how large are the data sets likely to be?*

3. *Who is likely to be in possession of these data sets or to be a gatekeeper to those who have the data available?*

4. *Are there established databases which will provide suitable data for the researcher?*

5. *How will it be possible to obtain access to the appropriate informants and how could they be motivated to provide the data?*

6. *How will it be possible to check the veracity of the data?*

7. *How will the data be collected and how long might this process take?*

8. *What type of triangulation will be available to the researcher?*

9. *What type of measuring instrument will be required?*

10. *How can the measuring instrument be validated?*

Few researchers will be able to answer these questions in any complete way as they set out on their research journey. However they are useful to ask as they provide some guidance as to what the data issues are likely to become as the research proceeds.

There should always be an element of opportunism in data collection and if the researcher comes across a new source of data which was not previously documented in the research proposal or research protocol then when appropriate this should be included in the research. Of course care has to be taken that the ethics protocol is not been disregarded.

Research questions do change and this will normally require different or additional data to be obtained and researchers need to be adequately flexible to rethink the answers to some of the above questions to accommodate this.

4.17 Summary and conclusion

This chapter/paper addresses a wide ranging topic which has not received the attention it demands. It is truly surprising how little thought is given to the definition and the issue of data and how often it is not incorporated in the research proposal and protocol. When academics are asked to define data they can find this difficult. The chapter/ paper has focused deliberately on issues related to the nature of data and does not provide an in depth discussion on how data is processed which it is hoped will evolve through further discussion.

Data is an issue of critical importance to academic researchers. It is in one form or another an integral part of academic research. The nature and general characteristics of data are seldom addressed as issues in their own right by academics, either in research papers or in books. This chapter/paper opens a discussion on the nature of data, the different forms it takes and how academic researchers should think about its acquisition etc. There is no suggestion that these objectives could be comprehensively achieved in one paper.

The issue of the definition of data indicates the complex nature of this subject as do the discussions on the use of data in research. The terms hard and soft data are seen as perhaps a misnomer. Data integrity is essential if it is to be used in academic research and researchers need to be aware of the confirmatory bias which can impact data integrity. Researchers need to be continually vigilant concerning these issues and be prepared to be critical of their own ideas and to change course when necessary.

This paper has not addressed issues related to the many methods available for data collection and or data analysis. It has not explored the data techniques surrounding questionnaires or interviews or the results of experiments. These topics are well discussed in the literature and do not require treatment in this paper. What the paper achieves is to describe the landscape in which data exists and should be considered and a checklist is provided to help with this. The issues discussed here are the first step in being able to understand what data is available, how it may be obtained and how it should be managed.

In general academic researchers do not spend enough time thinking about the issues described in this paper and it is hoped that by reading this work and reflecting on it researchers will give the question of data more attention.

> *In recent years universities have become much more demanding with regards to how they authorise research which has any involvement with living people or animals. In the social sciences an Ethics Protocol is nearly always necessary. Although this can only be regarded as a "good thing" it has made research more complex and some universities have encountered contradiction in the types of requirements they now seek from their researchers. The flexibility which Grounded Theorists has traditionally required to follow appropriate data is not always satisfactory to Ethics Committees. Some universities have asked researchers to obtain Ethics clearance a number of times during a research project when new data opportunities have arisen.*

5
Pragmatism

Confusing pragmatism with being pragmatic is an error.

A discussion of Grounded Theory's philosophical underpinning is often neglected and this is regrettable as it can lead to problems when the research findings are being justified.

It is normally well understood that every research method is based on a philosophical stance towards the world and our ability to understand it. It is for this reason that researchers claim that it is important to take an ontological and epistemological position which will in turn lead to an appropriate research methodology. It is difficult to argue against this. It should be stated, however, that many novice researchers find ontology and epistemology daunting.

If it is possible to point to a single most important philosophical influence on Grounded Theory some research philosophers would say that it was pragmatism. Of course pragmatism is not the only strong influencer but it is so important to Grounded Theory that it needs special consideration.

Grounded Theory eschews high cultural or formal philosophical explanations of knowledge and how to justify the findings of research. There is no attempt to find a teleological or a deontological position for the research. The justification is much more practical. If the findings of the research deliver something of value to someone then they are *right* in some sense. Furthermore pragmatism is keenly aware of the contingent nature of knowledge. As we simply do not know what we do not know and can never know what we do not know it is not possible to be dogmatic about our knowledge. Our conceptual maps and our conceptual capacity could at any time be turned on their heads and thus researchers require a degree of humility and pragmatism makes this quite clear.

Without an understanding of pragmatism the justification of the Grounded Theory method could be seen as rather shakey. Through the pragmatist lens Grounded Theory performs a necessary function in the everlasting pursuit of knowledge and for this social scientists are grateful.

However pragmatism does have its detractors and researchers need to be aware of what these arguments are. But at the end of the day the body of knowledge which we call social science has a strong stripe of pragmatism running through it.

This paper explores a number of different issues related to pragmatism as it may be used by a Grounded Theorist and was originally written as an occasional paper by Dan Remenyi.

Pragmatism

Pragmatism asks its usual question. "Grant an idea or be-
lief to be true," it says, "what concrete difference will its
being true make in anyone's actual life? How will the truth
be realized? What experiences will be different from those
which would be obtained if the belief were false? What, in
short, is the truth's cash-value in experiential terms?
— William James

5.1   Introduction

Pragmatism[1] is a philosophical stance which plays an important role
in much, if not most social science research. It is not so much a philosophical school as a philosophical trend in which the original, important and well known exponents are Charles Saunders Peirce, William James and John Dewey. The roots of
pragmatism are in the thinking of a number of 19th century American intellectuals,
who were seeking an alternative way of
understanding knowledge and its creation
or discovery to that of the metaphysical
traditions of European philosophers.
Although not many academic researchers
would claim to be Pragmatists, research

Academic research fol-
lowing a pragmatist ap-
proach does not mean
that the researcher is
automatically a
Grounded Theorist but
being a Grounded
Theorist researcher
does mean that the re-
searcher is informed by
pragmatism.

practices which flow from a Pragmatist point of view may be seen in

---

[1] Dewey (cited by Thayer, 1970) points out in his paper that Pragmatism has also
been know as Instrumentalism and Experimentalism.

the thinking and the writing of many social scientists. It is important to note that being pragmatic is not the same or similar to working and thinking as a Pragmatist.

5.2 Knowledge as 'coping strategies'

An important principle on which pragmatism is based is the notion that beliefs, ideas, theory and knowledge are 'coping strategies' (Menand 2002) which means that they only have value in so far as they provide a means of achieving objectives. Pragmatism is grounded in the work involved with practice and how this should be understood and is therefore essentially empirical in nature as well as relying on a strong rationalist bias. This type of research often draws on mixed methods in order to take advantage of the multiple lens opportunity that such an approach offers the researcher. Pragmatism does not recognise, or at least places little value in knowledge, for its own sake. Therefore Pragmatist research or inquiry (American spelling) has to have a practical orientation. It has to solve a problem.

> *Knowledge is indeed difficult to define and for the pragmatists knowledge is best acquired from doing and achieving some desired result. This of course goes against the old adage that something may be very well in theory but not any good in practice. Pragmatism may well have been what Lewin had in mind when he said, "There is nothing so practical as a good theory".*

Furthermore the result of any inquiry has to be understood within a social context. The important implications of understanding knowledge issues in a social context is that it is society or at least a relevant community, that recognises and therefore gives authority to knowledge, and that the practice of science evolves what is considered to be evidence. Wittgenstein (1969) recognised the need for knowledge to be understood as a social artifact when he pointed out that:

Knowledge is in the end based on acknowledgment.

In academic terms this means that the academy is the arbitrator and thus the academy offers the acknowledgement. For research degree candidates this is in the person of the examiners and through them the university degree awarding body.

Habermas (1993) supported the view that our understanding and acceptance of science is not static. Ideas resulting in new approaches to methodology and data are common, especially in the social sciences. Habermas commented that:

Now we think more tolerantly about what might count as science.

Gummersson is overstating the position but it is entirely correct to say that there are few people who can produce a robust definition of science. Scientists do not often take the time to define what it is that they do. Maybe it is similar to the famous George Bernard Shaw's quip, "Those who can do, those who can't teach ".

There are some scientists who have argued that scientific standards have in some senses been relaxed and that pseudoscience has acquired some degree of authenticity. But this is not what Habermas is referring to. There is an ongoing debate as to the nature of science and it is by no means near being settled. The older view that science was derived by the scientific method no longer seems appropriate as the scientific method has been shown to be more in the mind of the scientist than in his or her actual practice. Gummersson (2000) took a firm line on the inability to define science when he said:

> ….."scientists" who claim to know what science is are not scientists.

As an accomplished scientist himself, maybe Gummersson's statement appears to be somewhat dogmatic.

5.3 Theory and practice

Having accepted that knowledge and theory are 'coping strategies' for achieving objectives, the question which is raised is: *Where do these 'coping strategies' come from and how are they developed?* The Pragmatist believes that practice informs theory and that theory in turn informs practice. There is no gulf between the theory and practice. In fact the distinction is between informed practice and uninformed practice. The process of theory creation requires observation, reflection and formulation of concepts and the relations involved, i.e. the elements of the theory.

5.4 The literature

From an academic point of view the above notion of the creation of knowledge from observation, reflection and formulation of concepts and the relations is not adequate. Academe requires or perhaps prefers that knowledge creation and theory development be part of a historical process of incremental understanding. It is for this reason that a literature review is such a central element in the academic research process. Researchers are expected to use their inquiry to extend the current body of knowledge.

If the literature does not reveal an adequate platform of previous knowledge on which to develop new theory the research may use Grounded Theory to create a new theory. Although this is a well accepted approach it is generally regarded as more labour intensive than other research routes and it does not detract from the high level model of research practice of observation, reflection and formulation of concepts and the relations which drive the Pragmatist approach.

5.5 Pragmatism and Perspectivism

As Pragmatism reflects the importance of social context and the fact that the accepted view of science changes, it is sometimes seen as a form of Perspectivism. This is a philosophical point of view developed by Nietzsche (1967) which asserts that knowledge needs to be understood in terms of its context and especially the particular perspectives of the knowers. Nietzsche did not argue that all perspectives were equal, but rather that there is no one privileged route or understanding which can transcend cultural and personal perspectives. Nietzsche famously remarked, *"There are no facts, only interpretations"* and for this reason he has been seen as a comfort to extreme elements such as those who would deny the occurrence of the Jewish Holocaust.

The outright refusal to recognise the existence of any 'facts' is clearly problematic, but Nietzsche's remark is useful in that

The expression "There are no facts, only interpretations" can be uncomfortable, but it is useful in reminding us that there are different points of view and sometimes these can be antagonistic to one another. As academic researchers it behooves us to be aware of the different points of view and to be able to take them into account when we create our scholarly arguments. This is an important application of reflection in research.

129

it points out that what is considered a fact is often highly influenced by perspective and there are always multiple perspectives. However the expression *"There are no facts, only interpretations"* may be seen as little more than a rewording of the statement in Shakespeare's Hamlet[2], *"Nothing is good or bad but the thinking makes it so"* and Shakespeare does not attract this sort of criticism. Nietzsche argues that different understandings need to be aired and in so doing reach a consensus. This leads to a realization that there is no absolute knowledge, but rather knowledge which has been arrived at by human consensus[3]. This in turn suggests that knowledge is always a work-in-progress as new perspectives will inevitably arise over time and will cause knowledge to evolve. Babbie and Mouton (2001) explored this issue further when they wrote:

There is no such thing as an instant verification of a hypothesis or a theory. Even when a scientific community accepts certain points of view, hypotheses or theories, as valid and plausible, the acceptance is based on the best available evidence at a given point in time … If we were to accept a particular point of view as 'certain' or 'infallible' we are in fact saying that no amount of new evidence can ever lead us to change our belief. Such a view is not only obviously false, but clearly makes a mockery of the whole

---

[2] *Hamlet Act 2, scene 2, 239–251*
[3] In TS Eliot's poem Macavity-The Mystery Cat there is an interesting comment about the cat, "He's broken every human law. He breaks the law of gravity." Eliot understood that the Law of Gravity was nothing more than the human way of understanding a physical phenomenon which inter alia caused things such as cats, to fall.

scientific enterprise. The commitment to true and valid knowledge is, therefore, not a search for infallible and absolute knowledge.

The open mindedness needed for science is sometimes referred to as Fallibilism and is taken for granted today by many academic researchers. Fallibilism is an understanding that there is no guarantee that any scientific finding is immune to flaws and that it is quite possible that even the most accepted laws could be seen as being wrong tomorrow. The fact that it was believed for millennia that the Earth was flat and the Sun moved about it are examples of well-accepted falsehoods. Bishop James Ussher declared in 1685 that the Earth was created at 9.00 am on Monday[4], 23 October, 4004 BCE and this was accepted by many as being factually correct for years. In early Victorian times it was believed by some people that if a human body travelled at a speed in excess of 50 miles an hour their internal organs would be irreparably damaged.

Thus it is not possible to say which of our currently held beliefs will be regarded in the future in the same light as the above misconception.

5.6   Different lens and commonality of experiences

Pragmatism emphasises that there is no way of seeing the world other than through human eyes. It is not possible to have a greater than human understanding and human understanding is quintessen-

---

[4] There is some debate about the actual time that it was said the work started and it may have been late on the Sunday night.

tially personal. Understanding is always a product of the experiences, education and the agenda of the person concerned. For this reason it is sometimes said that each individual has his or her own reality. However the word reality can be confusing and thus it may be better to assert that each individual has his or her own personal lens through which the world is viewed. If this is the case then the question arises, *How can any agreement be reached about beliefs, ideas, theory and knowledge?* The answer to this question lies in the fact that although individuals are quite different in their understanding there is nonetheless an important level of commonality in the human experience. For example, human objectives around the world may differ in detail, but fundamentally these objectives are about how to live a satisfactory life, normally in a community, with due attention to the environment in its broadest sense. There are distinctly different codes of practice, traditions and laws which direct people to achieve this end, but most of them have common themes, employing rewards and punishment where it is deemed appropriate. Although commonality is not universal it is sufficient to allow human activity such as scientific enquiry to proceed, albeit cautiously. In this respect all that a scholar can appeal to are understandings acquired from experience, from reflection and the power of rational argument.

5.7   No appeal to a higher source of authority

There is in Pragmatism no appeal to a higher source of authority. Pragmatism is not attractive to those who are interested in classical philosophical discourse which inter alia appeals to Natural Laws or to Platonic forms or some other metaphysical construction and for this reason it has been strongly opposed by a number of important scholars. It shows no interest in any aspect of the Laws of Nature or whether the researcher has taken a Poperian or any other philosophical approach or perspective. Pragmatism is a consequentialist concept which asks, *What is the result or outcome of a particular course of action or what does it mean to hold such a point of view?* Some researchers abbreviate this with the abrupt question (Bryant 2009), *So what?* To answer questions like these, a Pragmatist looks at the detail of the results and makes a judgement in terms of the de facto situation. This can be attractive or a comfort to some, but an anathema to others who are looking to some greater authority for authentication. These concerns and objections to Pragmatism should not be entirely dismissed as there is a potential tendency for Pragmatists to take a somewhat WYSIWYG (What You See Is What You Get) approach to the world.

In understanding the role of Pragmatism in research the starting point is to recognise that research is a social activity and therefore has to be conducted with due cognisance of the relevance of the research problem and question, the rigour of the research process and the value, both in practice and in the theory, of the findings and results.

5.8   Research questions from practical problems

Whereas other academic research approaches would not necessarily welcome a practical problem as the starting point, pragmatism would see this as an almost essential prerequisite. Practical problems are found in real i.e. social situations, and a Pragmatist will spell these out in explaining the rationale for the research. It can be seen that what is now generally referred to as Mode 2 research (Gibbons 1994) by academics in business and management studies has a strong

Pragmatist orientation. Pragmatism is not suitable for pure research, i.e. research without a clear application, and therefore is an approach for applied or problem solving research. Nonetheless it is a mistake to suggest that the role of theory is in any way minimised by Pragmatists. Pragmatists use problem solving thinking in order to obtain an understanding of the theory behind the situations the research addresses.

5.9 The research process

Having established a suitable research problem, the Pragmatist will pursue an exhaustive literature review in order to understand the intellectual context of the research problem and to assist in the articulation of the research question. Because the research problem has been found in practice some researchers are tempted to feel that the literature review is of less importance. This is completely untrue and sometimes the practical nature of the research problem requires the researcher to have an even greater knowledge of issues surrounding the topic.

Moving forward with the research question the Pragmatist will, like other academic researchers, develop a research design which will be used to acquire appropriate data, manage it, and analyse it.

> *The implication for Grounded Theorists is that the closer they can get to being involved in the situation the better. This connects well with Lewin's action learning ideas and this leads to the idea that action research can be useful. Of course it is not always possible to undertake an action research program as a special type of access is required for this.*

The research design will often involve case studies, participant observer or action research as these methodological options are more likely to give the researcher fuller and closer access to the entity or entities participating in the research. Pragmatists argue sometimes that knowledge obtained as a spectator is not as useful as knowledge acquired by some form of hands-on experience.

Action research is thus preferred and if the researcher is employed in the organisation of interest then his or her employer could offer a suitable choice as a research site. It may be advantageous if the researcher is already known to some of the individuals who would participate in an action research project. Of course, conducting a research project in an employer's organisation does present challenges with regard to bias and the possible unwelcome intervention of colleagues, and the researcher will have to be careful that this does not happen. Pragmatism is agnostic with regard to whether the researcher has a preference for quantitative or qualitative data. In action research it is likely that both data types will be used and thus if the research is conducted thoughtfully the researcher will probably want to claim that a mixed methods approach has been used (Creswell et al. 2003). A mixed method approach is intellectually attractive but it is also quite demanding and it has not received as much acceptance as it was originally thought it would.

Pragmatists use the term competent inquiry and by this they mean that the execution of the research design has to ensure that an adequate degree of understanding and rigour, judged by the community in which the researcher is working, is applied to all the research processes. It will be because the inquiry has been competent that the researcher will be entitled to make what Pragmatists call a warranted assertion that knowledge had been produced. Pragmatists sometimes use the term warranted assertion as a substitute for knowledge as they feel that there is a degree of ambiguity associated with the word knowledge. It is important to understand that there is no calculus for the establishment of whether something is a warranted assertion. A warranted assertion is established by the power of the argument or the rhetoric on which it is based.

5.10 Ceteris paribus

Observing a Pragmatist during these phases of the research may not reveal differences to the way in which other researchers who are engaged in action research or case studies will be working. However the Pragmatist will be perhaps more keenly aware of any shortcomings in the research process and the limitations in the theory which may result from the research. Theoretical understanding is always subject to variations, especially in the social sciences as theorists cannot be certain that all the variables in the situation are incorporated in the research. This phenomenon is well understood and is accounted for in the assumption which underpins most academic social science, which is ceteris paribus i.e. all other things being equal. Feyerabend (1993) although referring more to physical and life sciences was clearly aware of this when he wrote:

> ...... we may start by pointing out that no single theory ever agrees with all the known facts in its domain. And the trouble is not created by rumours, or by the results of sloppy procedure. It is by experiment and measurement of the highest precision and reliability.

An awareness of this issue is central to all academic research but especially to that conducted in the social sciences. However it is seldom explicitly addressed due to the focus which researchers usually apply to their research question. Pragmatism requires the limitations of research to be understood and to be kept in the forefront of the researchers' minds. Feynman (1995) gave a particularly good explanation of these limitations when he wrote:

> *Each piece, or part, of the whole of nature is always merely an approximation to the complete truth, or the complete truth so far as we know it. In fact, everything we know is only some kind of approximation, because we know that we do not know all the laws as yet. Therefore, things must be learned only to be unlearned again or, more likely, to be corrected.*

It is the interpretation of the finding which comes from the analysis of the data that is one of the most clearly distinguishing features of the Pragmatist approach. In this paradigm, for the findings to be regarded as knowledge and thus of value it is necessary for the researcher to show that he or she is able to produce insights which can be used to make a difference to the understanding of the problem and that these have some value in practice. This use in practice is a central issue. In this respect a social or communal approach is required and thus the researcher needs to include opinions from his or her community of practitioners in support of his or her findings and ultimately his or her warranted assertions. This is in practice equivalent to the more commonly encountered notion of validity. The research outcome could be referred to as having established what Glaser and Strauss (1967) refer to as Substantive Theory and thus it would not be able to claim a significant degree of generalisability. A Pragmatist would not recognise any value in the issue or replicability as they argue that every situation will be different

(Wheatley 1992) and that therefore replicability is not possible. This does not mean the findings have no value. It does mean that claims made for the findings have to be expressed with considerable care and special attention needs to be given if the researcher wishes to assert that the findings have value beyond the situation researched.

5.11 Knowledge is always a work-in-progress

For the Pragmatist the creation of knowledge is always a work-in-progress. This is due to the complexity of the world which the researcher tries to understand as well as the social nature of the discovery process. Some researchers argue that research should be seen as an activity which displays many of the attributes of biological evolution in which better theories survive and poor theories become extinct. Russell (1960) suggested that pragmatism should be thought of as evolutionary. As a philosopher Russell was looking for certainty with regard to the outcomes of research and consequently had reservations concerning Pragmatism's open-ended view regarding the creation of knowledge. However today there is increasing awareness of the work-in-progress view of knowledge and thus there is considerable support for the view that there is no final position from which absolute knowledge may be claimed. Checkland (1986) shows an awareness of this when he stated:

> *Obviously the work is not finished, and can never be finished. There are no absolute positions to be reached in the attempt by men (sic) to understand the world in which they find themselves: new experience may in the future refute present conjectures. So the work itself must be regarded as an on-going system of a particular kind: A*

learning system which will continue to develop ideas, to test them out in practice, and to learn from the experience gained.

This idea has become accepted to the point where it is an article of faith among many scientists although it is said from time to time that an experiment has proved once and for all that a certain proposition is 'true'. This of course is simply 'bad' science. It would be much more acceptable to say that the scientific evidence suggests that the proposition be *regarded* as 'true' (this is intrinsically a difficult word and it should be avoided) pro tem or for the time being.

5.12 Important areas of concern

As mentioned above there are a number of important areas of concern when reviewing the Pragmatist research paradigm and these include:

1. The possible definitions of usefulness in practice and;
2. the degree of usefulness;
3. The expected duration of the usefulness of knowledge;
4. The connection between the usefulness of knowledge and any moral considerations.

Usefulness in practice is sometimes a subjective notion. It would always require careful definition and it could not be expected that all members of a community would be in full agreement with such a definition. Then there is the problem of degree of usefulness. *What degree of usefulness is necessary for the outcome of the research to be awarded such designation?* Academic research offers an additional complication. Sometimes a research project does not result in the research question being answered in such a way that a solution is found. The researcher may find that, at least to some extent, the problem is intractable and this may not satisfy his or her community. When this happens a difficult question is raised: *Would a Pragmatist approach regard this research and its outcome as having failed?* Traditionally an academic research project may be considered successful if it leads to the researcher being able to claim that he or she has added something of value to the body of theoretical knowledge and this could be in certain circumstances that the research question was not in its current form answerable. It may be that the research has led to a greater understanding of the topic which in turn allows the researcher to frame better questions. This could be academically acceptable, but it would not necessarily solve any problem and thus would not be likely to satisfy the needs of a practitioner facing the original problem.

Given that the problem of the definitions of usefulness and the degree of usefulness can be to some extent resolved, then there is the issue of the timeframe. What is useful today may not be useful tomorrow. In fast moving areas usefulness may be a fleeting condition and thus would findings like these really warrant being regarded as an academic contribution? From a Pragmatist point of view change is recognised as inevitable and thus if the solution is not useful tomorrow this will represent another opportunity for an additional competent inquiry.

Some critics of Pragmatism have been uncomfortable about the fact that Pragmatism does not address the issue of whether a useful course of action should be influenced by any moral considerations. This is of course correct but such critics ignore the fact that scientific research in general does not attempt to claim any moral stance. Many findings in all branches of research may be used for good or for evil, on the assumption that it would be possible to define 'good' and 'evil'. Any attempt at such definitions would be highly problematic and would no doubt attract the attention of extreme points of view. There is discomfort among researchers concerning nuclear physics being used to create weapons, genetics to create modified crops, computer scientists concerning how personal privacy can be affected by large databases, to mention only a few examples of controversial issues. However scientists are often relatively silent on these issues, leaving moral comment to others. There is a long tradition of authors taking up the subject of amoral science ranging from Mary Shelley's Frankenstein to Michael Crichton's Jurassic Park. In Jurassic Park the scientist character remarks ironically that they were too busy working on the 'how' question to have time to ask the 'if they should' question.

In business and management studies research and scholarship Ghoshal (2005) points out that there are a number of important moral issues which have had adverse consequences and which need considerable and urgent attention. Ghoshal believes that the morally neutral stance which business and management studies have attempted to take has led to management abuse in organisations such as Enron and Worldcom. Ghoshal was writing before the 2007 financial meltdown which provided further evidence of his contention and which led to the world wide recession.

5.13 A possible paradox

From the above it may be discerned that within the Pragmatist framework of thinking there is a potential paradox. As Pragmatists understand that knowledge is socially determined and cannot be regarded as complete (it is always work-in-progress) there is a predilection to see this approach to academic research as suited to a social constructivist approach. After all, one of the basic tenets of social constructivism is that there is not one privileged way of knowing the world and this would seem to be an approach appropriate to pragmatism. Social constructivism is often employed where there are issues about the correspondence of cause and effect and about the quantification of the variables under consideration.

But Pragmatism is interested in practical or useful outcomes to the research and sometimes this aspect of the research is a challenge. It has already been noted that there are difficulties with the definition of the word 'usefulness' and with the notion of degrees of usefulness. The same applies to notions of prestige, social responsibility, satisfaction, degree of corporate governance, to mention only a few issues. However, academic research does not only address such issues where there are measurement challenges. An academic research project could address and produce a new theory, which when applied could be demonstrated to have generated a specific level of new business, which could be measured quantitatively and thus be seen as a result that has direct realist implications.

5.14 Pragmatism and Neopragmatism

The success of Pragmatism can be assessed not by how many researchers regard themselves as Pragmatists but rather by how the fundamental tenets of Pragmatism have become taken for granted principles of research. Issues such as the uncertainty of research findings, the work-in-progress nature of knowledge, the recognition of the social nature of research, the need to couple theory and practice, have all been taken over by main stream researchers especially in business and management studies.

In the past few decades there has been a resurgence in interest in Pragmatists and these individuals are now referred to as Neopragmatists. Neopragmatism is a post modern version of the ideas of Sanders, James and Dewey. It places greater emphasis on linguistics, epistemology and hermeneutics while still drawing on the works of the founders. It is of greater interest in the general philosophical discourse than it is to the day-to-day operation of research methodology.

5.15 Getting going as a Pragmatist

Pragmatism is not a methodology and thus there is no practical framework detailing the steps required to produce Pragmatist informed research. The following represents some of the issues which will have to be addressed by a researcher interested in producing research directly influenced by Pragmatism.

A Pragmatist's philosophical stance requires:

1. The research question needs to be selected so that if it is answered it will solve a problem which will be of direct use to the researcher's community. Remember that the research question still needs to be academically authenticated by reference to the extant literature.

2. The research design needs to seek an environment in which the practical value of the research can be demonstrated. This will lend creditability to the research process and will probably involve action research or participant-observer procedures.
3. The use of multi method type procedures for data collection, management and interpretation.
4. The validation of the findings in terms of demonstrated utility or usefulness to the researcher's community.

None of these tasks are trivial and researchers who wish to pursue this approach to their work need to proceed with circumspection and care. On the other hand it needs to be said that Mode 2 research which is increasingly accepted as good practice in business and management research follows the same basic philosophical approach as Pragmatism. It is likely that Pragmatism will play an increasing role in this type of research in the future.

5.16 Summary and Conclusion

In Pragmatism from a purely epistemological point of view, knowledge is a coping strategy which is formulated from the experience of those working in the everyday environment. The worker behaviour/understanding is observed, reflected upon and formulated in such a way that it becomes a theory. Created like this it is always a substantive theory, in the Grounded Theory sense. So, epistemologically knowledge created through the Pragmatist lens is first and foremost empirical. Because there will be at some level an interaction between the researcher and the observed, the research process is always social. This means that there will inevitably be some degree of social constructivism about research undertaken in this way. The issues underpinning social constructivism are even more prominent when it comes to deciding whether the findings are 'useful'.

So in short, Pragmatism is not an epistemology. It is a philosophical stance which guides the researcher in his or her methodological choices. Not a methodology itself, Pragmatism informs the research process and produces firm guidelines on evaluating the results of re-search. The Pragmatist will always acknowledge the uncertainty as-sociated with all the processes of the research and will regard the outcomes as work-in-progress.

In conclusion, Pragmatists argue that their philosophical stance leads to more appropriate research being conducted in situations which can deliver benefits in a more direct way. The understanding of the nature of knowledge creation whereby practice informs theory and theory informs practice has been of considerable value in under-standing epistemological issues. In so doing the researcher needs to minimise the distance between him or herself and the organisation and the individuals participating in the research. The notion of the researcher being a disinterested observer is a fallacy or perhaps even a fantasy. The research question, the research process and the re-search findings are understood to be drawn on social awareness, ac-tivities and requirements in which the researcher will be highly in-volved. This will produce challenges relating to bias, but researchers should be capable of successfully meeting these challenges. Pragma-tism clearly articulates the fragile nature of knowledge and under-standing as it firmly rejects any pretence to permanence in this re-spect.

It needs to be stated that there are shortcomings which can arise in the use of Pragmatism in academic research and researchers should be cognisant of these. However, many of the ideas raised by the Pragmatist have been incorporated into academic research practice and this has generally been beneficial. On a practical level it is perhaps the accommodation of the importance of the individual's ability to make a judgement and assess the value of a situation which makes Pragmatism interesting to researchers and scholars.

Some researchers find pragmatism unsatisfactory and they argue that there needs to be more to knowledge than a copying strategy as suggested in this paper. Taking this point of view leads researchers into a philosophical debate which ends with there being a number of different points of view and not much concordance. Although research often thrives on disagreement there needs to be some general understanding of the parameters within which competent academic research may be conducted and some social science researchers find pragmatism useful in this respect.

As mentioned earlier pragmatists are not necessarily Grounded Theorists, but Grounded Theorists are informed by pragmatism and to the extent that this is the case Grounded Theory may be seen to directly deliver utility.

In the end it is the task of the researcher to pick and choose his or her toolset from the array offered by Grounded Theory in order to be able to make a convincing scholarly claim that he or she has added something of value to the body of theoretical knowledge.

6
Grounded Theory at Work

A convincing argument is the ultimate objective of the researcher.

This chapter describes some of the issues which in practice require attention when taking a Grounded Theory approach to academic research. Of course every researcher will interpret his or her own specific needs and therefore being too prescriptive about the method may sometimes not be helpful.

Creating a flow diagram for Grounded Theory can be misleading because of the iterative nature of the approach. It is therefore probably better to state the main activities or processes involved and to emphasise that the researcher needs to stay fully immersed in these processes until an acceptable theory emerges.

An acceptable theory is one that the community in which the researcher is working acknowledges is a useful explanation of the phenomenon being studied. The explanation could be relatively superficial or it could be comprehensive, but neither of these directly affects the issue of its acceptability.

6.1 Paraphernalia of Grounded Theory

The term paraphernalia, used in the diagram below, refers to the different activities which contribute to a Grounded Theory approach to research. Although there is the opportunity for researchers to design their own research program all the activities suggested in Figure 1 will probably be present to some extent in many qualitative research programs. The activity which could be completely absent from a qualitative research program would be coding, if a hermeneutic approach was being taken. Although there is some natural order in which these activities take place, in that data has to be acquired before it can be coded and categories are developed after coding and sorting, many of these activities happen in parallel or are repeated a number of times. In fact one of the major differentiating features of Grounded Theory is that the researcher stays involved with the data and repeats activities like sorting and conceptual mapping until he or she is satisfied that a useful theory has emerged.

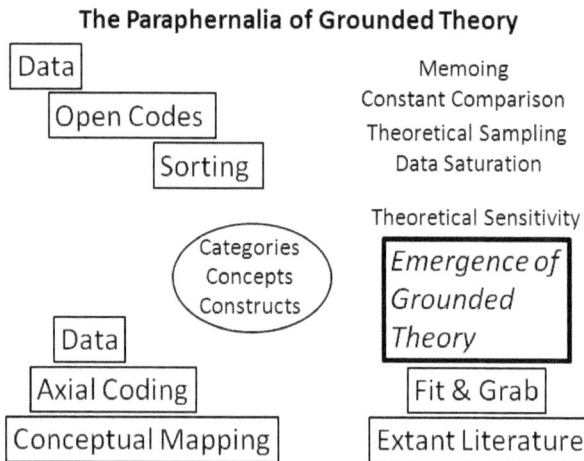

The Paraphernalia of Grounded Theory

Data	Memoing
Open Codes	Constant Comparison
	Theoretical Sampling
Sorting	Data Saturation

Theoretical Sensitivity

Categories Concepts Constructs

Emergence of Grounded Theory

Data	
Axial Coding	Fit & Grab
Conceptual Mapping	Extant Literature

Figure 6.1: The Paraphernalia of Grounded Theory

It has been pointed out that there are many different activities which the researcher has to think about within the Grounded Theory method and in this respect Figure 1 could be seen as a check list.

6.2 The research question

Before the data collection process can commence in a serious way a research question has to be established. Without a research question a researcher can only make relatively superficial enquiries about the topic of interest. It is not necessary to have finally settled on a research question before beginning to collect potentially useful data. However it needs to be remembered that most universities require specific permission to be granted to academic researchers before they are allowed to commence an academic research project and the Ethics Committees which grant such permissions will inevitably require not only a research question to be settled, but also a detailed research protocol to have been developed. Universities require researchers to closely comply with these protocols and thus the flexibility which has been a characteristic of Grounded Theory has had to be in some cases curtailed. Whereas in the past researchers could take the view that they would *follow the data* wherever it might lead, this is no longer permitted by university ethics practice. If a researcher wants to pursue a new line of enquiry which has arisen during the research he or she will probably have to make another application to the Ethics Committee to be allowed to do this. Of course different universities apply their Ethics rules differently and all researchers should now be sure to understand what is required by their own institution.

6.3   The dataist approach

Employing a dataist[1] approach to research a Grounded Theory researcher will collect data continuously throughout the project and
reflect on the data as it is collected and this plays an important role
in blurring the distinction between data collection and data analysis.
Although the researcher will be open to data from a variety of
sources Grounded Theory often focuses on interviews and data obtained through this question and answer process can be supple-
mented by written data of various sorts. The researcher finds that
his or her thinking will be developing throughout the project as a result of this data acquisition. The distinction between data collection
and data analysis will blur. While the data collected is being processed the researcher is required to record his or her reflection in
memos which will become an important element in the final theory
development. This means that the researcher's learning process,
which is central to all academic research, needs to be active throughout the whole research project. The aim is continuous learning on the
part of the researcher, acknowledging that at any point a new insight
might become apparent.

6.4   Formal data analysis

Grounded Theory also requires formal data analysis and this is performed through data coding, sorting, grouping and category formulation. This type of approach to data analysis is sometimes referred to

---

[1] All empirical research is essentially dataist. The term is used here to suggest that
Grounded Theory is generally more data driven than most other research.

as Content Analysis[2]. Some researchers will code after each data collection episode while others will code on a periodic basis. Because of the continuous learning aspect of the research and the need for continual comparison the formal data analysis needs to be performed regularly.

Data coding is a central feature of a Grounded Theory project and the selection of codes is most important as the chosen codes have a direct influence on the outcome or findings of the research. Researchers need to be flexible and change codes if the situation suggests this course of action. Sometimes it is useful to ask other researchers to code a part of the transcript to obtain an impression of how another mind sees the data. The coding is only a step, although an important one in moving from data to categories and thus it has to be seen as a way of facilitating a greater understanding of the phenomenon being studied.

---

[2] It is inevitable that there will be researchers who will take issue with the term *formal data analysis* and how it is used in this chapter. There will also be those who consider that the Grounded Theory approach to data coding, sorting and grouping does not employ Content Analysis to its fullest extent and that in the hands of Grounded Theorists this coding etc., should not be called Content Analysis. Recently a student raised the following question, *"If the transcripts are not coded but analysed by a form of hermeneutic analysis can the method used be referred to as Grounded Theory?"* In answering this question it is important to say that labels in research methodology are important, but correctly labelling one's research is not the only issue and thus whether the research is labelled Grounded Theory or not may be of little consequence. It is rather unlikely that the original authors of grounded Theory would be delighted to find that hermeneutic had replaced coding etc. But there are those today who argue for a more constructivist attitude to the data analysis and they are not concerned that their research has been weakened in any way by this approach.

Coding alone is not sufficient to understand the data. The codes have to be summarised and then they may be sorted in different ways. These activities have to be accompanied by considerable reflection and the meaning which the researcher will eventually give to this data will emerge slowly. Rushing this aspect of the research is a mistake.

Open coding is the Grounded Theory language for first cycle coding and inevitably there will be a need for a second round of coding, i.e. grouping the codes into supra-codes or even into categories. This second cycle coding is referred to as axial coding. It is can be helpful to display supra-codes or categories as perceptual maps which indicate potential relationships between the ideas being generated.

6.5   Data saturation

Dataists need to aware that they should collect an adequate amount of data but no more. It is tempting to overdo the amount of data collected which will slow down the research and can even distort the findings. Data asphyxiation needs to be avoided and thus the question of data saturation is important. Data saturation occurs when the researcher finds that additional data collected are not revealing any new insights into the phenomenon being studied and thus data collection may be terminated. The question of from whom data should be collected is largely addressed for Grounded Theory researchers through the concept of theoretical sampling. Theoretical sampling focuses on seeking out additional data sources which are likely to supply data that will contribute to a fuller understanding of the categories being developed. Thus it is a form of purposeful sampling whereby knowledgeable informants are sought. There is no place for random sampling in the Grounded Theory world.

6.6   Not all data are equal

Before leaving the topic of data collection it is worth mentioning that not all data are equal and the researcher needs to keep this in mind. Ideally researchers would like to have access to knowledgeable informants who are at the centre of the processes or phenomenon being studied, but access to such people can be difficult. In practice researchers often have to compromise with regards to this. It is always a question of judgement as to which data the researcher regards as suitable, but whatever decision is made the researcher needs to be able to justify it. In the same way what is heard from certain knowledgeable informants will make a greater impression on the researcher and this needs to be recognised. The inequality in data is one of the great challenges of qualitative research.

The purpose of the data collection and analysis is to produce categories or concepts or constructs as a stepping stone towards theory development. Recognising that these categories are the building blocks of the theory being developed Grounded Theorists require that the categories be as fully understood as concepts in their own right before they are incorporated into a theory. This is the main motivation for theoretical sampling and is referred to as populating the characteristics of the categories.

6.7   The most creative part of the research

The production of well understood categories is not the same as theory development. Theory development is the next stage and requires the researcher to take an additional step in envisioning how the categories could work together and produce an outcome. It is the description of how the categories work together and the outcome they produce which constitutes the theory. This is the most creative part of the research process and to a large extent it draws on the imagination of the researcher. But it is not an ill informed imagination which is required. The researcher needs to have a degree of theoretical sensitivity. Theoretic sensitivity is developed through learning, experience and reflection. If the researcher is well versed in a number of theories in his or her field of study then he or she is likely to be more theoretically sensitive. This is one of the reasons why Grounded Theory is more suited to mature researchers than to traditional undergraduates.

The way that theory finally emerges is particular to each researcher. Some researchers report that conceptual mapping is a major help to them. There are matrix techniques which can be used to draw attention to how different ideas are related and how they might impact on each other. Other researchers report that they reflect on the categories they have identified until they can create a list of possible theoretical explanations and then air them to associates and colleagues. At this point the dialectic becomes part of the theory development process and it encourages as well as fine tunes the researcher's original thoughts.

6.8   Substantive and formal theory

Although Grounded Theorists are primarily focused on producing substantive theory, they will often have the creation of formal theory as the ultimate objective in mind. There is no contradiction here as theory tends to evolve over time and if the theory is sound it will grow in its possible applications and even in its potential for generalisation. It is a generally believed a myth that Archimedes, Galileo and Newton had instantaneous realisations about the nature of the phenomena they were studying. Strong and useful theories improve as more evidence comes to light and understanding increases, while poor theories tend to wither or fade away as their weaknesses are exposed.

However there are two important issues which researchers have to bear in mind. The first of these is that any given theory represents the best explanation which can be produced at that time. As such it is normally imperfect in that there will usually be exceptions or anomalies. Feyerabend (1990) pointed out:

> ...... we may start by pointing out that no single theory ever agrees with all the known facts in its domain. And the trouble is not created by rumours, or by the results of sloppy procedure. It is by experiment and measurement of the highest precision and reliability.

The reason for these anomalies is usually attributed to the underpinning use of the philosophical assumptions embedded in certeris paribus and Ockham's razor.

The second issue relates to the ongoing nature of scientific discovery. Human endeavour to create knowledge is a process with no definitive end point. Knowledge creation has always been achieved one step at a time and every step has always been celebrated as a remarkable event. But as soon as some new knowledge has been at-tained, this knowledge itself usually opens up new horizons which indicate that there is more knowledge to be acquired. Checkland (1986) expressed the scientific attitude to the ongoing nature of research when he remarked:

> *Obviously the work is not finished, and can never be finished. There are no absolute positions to be reached in the attempt by men to understand the world in which they find themselves: new experience may in the future refute present conjectures. So the work itself must be regarded as an on-going system of a particular kind: A learning system which will continue to develop ideas, to test them out in practice, and to learn from the experience gained.*

Successful Grounded Theorists easily relate to this mindset. From an academic degree point of view the question is always how much of a contribution is required at any given point. This question is always challenging and the answer is a function of the circumstances of the research project. But it may be said with confidence that researchers should not be distracted by believing that they are seeking a final answer, but rather an improvement in the current level of understanding.

6.9   Fit and Grab

In the language of Grounded Theory the concepts discussed in the section above relate to fit. A theory is said to have fit if it adequately explains the data related to it and it always has to be remembered that even the most carefully developed theory will have exceptions or anomalies. On the question of grab, which is a word that does not have any academic ring to it (application might have been a better word to use in this context), theory which is recognised by the community as having some value and is employed in practice is regarded as having grab. Ultimately grab validates the research findings and can be seen as the pragmatist philosophical stance in action. It is difficult to over state the importance of grab. Pragmatism is underpinned by the proposition that if something works then in some sense it is right. This justifies the Grounded Theorists research method and allows the finding of the research to be referred to as warranted assertions.

If there has been no grab then the researcher has to ask him or herself what has been the value of the research project. As mentioned earlier there is little or no place for "pure" research in the world of the Grounded Theorist. But neither is there any reason why grab or application should take place instantly or even quickly after the new theory has been postulated. It can take time for theory to be recognised as useful and to be put into practice by the community. This is just one more example of the many paradoxes which are encountered in academic research practice. F Scot Fitzgerald[3] was correct when he said:

> *The test of a first-rate intelligence is the ability to hold two opposed ideas in mind at the same time and still retain the ability to function.*

---

[3] Published posthumously by New Directions in 1945 The Crack Up is an account of F Scot Fitzgerald's decline in the last years of his life.

6.10 The clean slate or tabula rasa

There is one other important element in Figure 1 which needs careful reflection and that is the issue of the extant literature. Grounded Theorists traditionally eschewed the academic literature because it was believed that being steeped in what others had thought would guide the researcher towards theory testing rather than theory development. There were also concerns about the confirmatory bias creeping into the work. Although there is some validity in this thinking it is now generally accepted that some familiarity with the literature cannot be avoided. The so called "clean slate" or tabula rasa approach is a fiction. What is more realistic is to state that the Grounded Theorist should not be unduly influenced by the extant literature. However such a statement opens up another area of controversy and possible paradox. It is now generally recognised that a researcher can only make sense of data if he or she perceives it through some sort of paradigmatic filter.

It is also important to note that theoretical sensitivity implies knowledge of theory in general and this most frequently requires knowledge of the extant literature.

Whatever the position taken on pre-research knowledge of the literature it is important that the emerged theory be validated in terms of the extant literature. The question is, *Do the Grounded Theory research findings resonate with what else is known about the field of study?* If they don't then it is important to know why. Answering this underpins the creditability of the findings of the research.

There is also the issue that for an academic degree the theory produced by the Grounded Theorist will have to be original and therefore it is essential that the researcher be aware of the thoughts of others.

In addition, in the Grounded Theory environment, the use of the extant literature is one method of moving from substantive to formal theory as is the acquisition of further data and new insights acquired by discourse with other researchers.

6.11 Summary

Making Grounded Theory work is not trivial. Although cognitively challenging until the concepts are reflected upon and understood, it is not intellectually mind breaking either. It is not a quick method. It requires a logical and careful approach which in general cannot be easily rushed. It does however in general lead to findings which may be regarded as emanating from competent enquiry.

Grounded Theory is appealing to research novices who are often looking for very specific guidance on how to conduct their research. More experienced researchers are often more eclectic and want to pick and choose the elements of this approach which suits them.

7

What Characterises Grounded Theory Research?

Grown up Grounded Theory has become more than the sum of its original parts.

7.1 Attributes of a Grounded Theory study

A question which is sometimes asked is what does a project that has been pursued using Grounded Theory look like and how does it differ from a non-Grounded Theory piece of research. To some researchers this is an important question, especially those who feel that they want to follow a well established prescribed route to their research. However this is not an easy question to answer and some method-ologists would say that it is not a question that should be asked as they would suggest that research methodology does not come in clearly labelled boxes and so there will always be fuzzy boundaries and overlaps between methods. However the reality is that this question does get asked, all too often.

The following key attributes suggest that a piece of research could be considered competent Grounded Theory based work:-

1. The research focuses on theory development;
2. There is a dataist attitude toward the field work and its un-derstanding;

3. The literature review is conducted with care to minimise any preconceptions;
4. The principle of theoretical sampling is applied.

It will be noted that it has not been explicitly stated that the research will be empirical and induction orientated as this is implied in the other points listed. The emphasis placed on these issues will differ but in a Grounded Theory there will be indications that all four of these dimensions have been carefully addressed.

7.2 The research focuses on theory development

At first glance this appears to be a factor which will be easy to assess and in many ways it is. Grounded Theory is not suitable if the objective of the research is to test an already established theory through some process of hypothesis testing. The differences between theory development and theory testing permeate right throughout the whole research process. In the first instance the type of research questions will differ. In theory development the research question essentially has the form of *What theory would be useful in understanding the phenomenon or phenomena under study?* This has a problem solving orientation in which the researcher is seeking a way of solving a challenge or obtaining a better understanding of an issue. In theory-testing the research question may have different formats, but they will be attempting to reject[1] an hypothesis. Thus the research question or perhaps better expressed the research problem

---

[1] It is a fundamental principle of research practice that an hypothesis cannot ever be proved. The researcher tries to reject it and if this is not achievable then the hypothesis is accepted pro tem.

will be stated in the form of a claim such as *Managers cannot be held responsible for performance if they are not provided with adequate resources to manage.*

Those involved with theory development and theory testing will have different attitudes towards what constitutes data; how it is acquired, analysed and interpreted. There are indeed important differences between how research conducted for the purposes of theory testing and theory development is approached. It is true to say that they are sometimes over stated as it is clear there are aspects of theory development which need to embrace some aspects of theory testing. After the data necessary for theory development has been acquired, transcribed, coded, sorted etc., possible theories may begin to emerge. Any one research situation could result in multiple possible theoretical explanations, which could be considered as candidate theories, and the researcher needs to consider these carefully and decide which is the mostly likely to provide the best or at least the preferred theoretical explanation[2]. Expressed differently, the researcher has to test the candidate theories and make a decision as to which one is offered as the research finding. This process of considering a range of possibilities has been referred to as abduction, but it is clearly similar if not exactly the same as induction performed in a relatively informal way (Birks and Mills 2011).

---

[2] The preferred theoretical explanation does not necessarily mean that it is the correct explanation, but rather that it is at that time the most likely.

Each theory speculation conceived by the researcher may be viewed as a candidate theory which has to be considered carefully before it can be either dismissed (rejected) or not dismissed (not rejected).

The sort of candidate theory testing required could be little more than the use of thought experiments, as these can be effective in choosing between alternative explanations. However in this respect the aphorism "two minds are better than one" applies and academic researchers will normally benefit from exposing their thinking to others and engaging in a dialectical discussion.

This theory development nature of Grounded Theory research also reflects the thinking of pragmatism in that, as mentioned above, the researcher is looking to solve a problem. The process of articulating multiple theoretical explanations that could facilitate the solving of the problem and selecting the one with the best explanatory capability is also pragmatism at work.

7.3 There is a dataist orientation

The dataist orientation involves examining and re-examining data which is regarded as the way of acquiring an understanding of the situation and answering the research question. This is done with constant comparison techniques in mind as it is believed that looking for patterns of similarity or difference leads to understanding. Humans have been described as pattern seeking creatures, which from a research point of view is most often an asset. However spurious patterns can and do arise and care has to be taken to avoid these.

It is important to note that the data should be examined several times as it can take some time for ideas to formulate or develop in the mind of the researcher. This is of course not exclusive to Grounded Theory, but other approaches to research would in general not place the same degree of emphasis on this close scrutiny of the data. In some respects other methods would take it for granted that the data was to be examined with great care.

Grounded Theorists recognise that in dealing with qualitative data there are difficulties in conceptualising data acquisition and data analysis as entirely different dimensions of a research project. There is no doubt that in many instances the researcher's understanding of qualitative data begins during the acquisition process. This is the result of engaging with knowledgeable informants and perceptive researchers will be learning from this experience.

Many Grounded Theorists propose data coding as the way of understanding the phenomenon being studied. It is this coding orientation which presupposes the counting of concepts that has mostly lead to some researchers referring to Grounded Theory as being positivistic in nature. However it is possible that data could be coded, and the resulting content analysis be used in a relatively non-positivistic way. As mentioned several times in previous chapters counting alone does not dictate a theory testing or cause and effect type orientation to the research.

Coding fractures the data in order to find issues, concepts, constructs or categories. Techniques of content analysis then help to set a context for conceptualising the importance of the various ideas and their relationship to one another. Re-examination of the data together with appropriate reflection leads to richer understandings of the categories. Finally the researcher has to integrate all these findings and produce a theory. Here theoretical sensitivity is required although, as mentioned above, the researcher is encouraged not to allow his or her thinking to be unduly influenced by preconceptions acquired from the extant literature.

Occasionally the question is raised, *Can Grounded Theory be conducted without data coding?* Perhaps another way of putting this question is *Can some sort of hermeneutic analysis be substituted for data coding in a Grounded Theory project?* In answering this question it is important to say that there is no organisation that controls the way in which humans describe their activities, and if someone decides that they wish to claim that their research has used Grounded Theory without employing coding then the only sanction would be the lack of acceptance of that categorisation within the community of scholars. It was clearly not the intention of Glaser and Strauss to include hermeneutic analysis in the method they described. But why should Glaser and Strauss have the last word in what constitutes Grounded Theory? Like most other aspects of human existence Grounded Theory principles are subject to change and to development and refinement and this has been pioneered by Bryant, Charmaz and others. There is no reason why the data cannot be fully understood using constructivist or interpretivist approaches and be used in producing competent Grounded Theory.

With Grounded Theory's strong emphasis on data it is somewhat surprising that there is seldom provided a comprehensive definition of data and the researcher is largely left to his or her own devices to establish what might be considered a useful definition with which to work. Glaser's remark "All is data" is simply unhelpful as there are often blind allies which do not lead anywhere. It may even be useful to think about anti-data, which could be defined as any apparent data the researcher finds, but which turns out to make his or her research journey more difficult.

7.4 The literature review is conducted with care to minimise any preconceptions

Academic research without a literature review is akin to having a ship without a rudder to guide it or indeed a ship without any sort of map to suggest the way in which the crew should plot a course to their destination. In academic research it has been traditional for researchers to show evidence of familiarity with the body of literature surrounding the topic being researched. It is hard to imagine competent academic research which does not build on the literature that is the written record of knowledge and which has been acquired over time on a particular topic. Academic research seeks to add new knowledge to what is already known and there is no other way of establishing what is already known than addressing the literature. However it is obvious that a researcher's thinking will be directed if not channelled by the literature and thus the challenge of seeing new issues or dimensions can be increased if the extant literature dominates the researcher's perspective.

By suggesting that the literature review be approached differently Grounded Theory attempts to minimise researcher preconceptions. There are a number of issues which make this difficult, but one of the characteristics of a Grounded Theory research project is that there will probably not be an exhaustive literature review completed before the empirical research begins. Grounded Theorists now argue that an exhaustive literature review should be conducted after the new theory has emerged (Urquhart 2013). The purpose of this is to ensure that it is at least to some extent original, i.e. it has not already been done and published elsewhere and that there is evidence of its importance from what is being said by other scholars.

The issue of how the researcher's cognitive map is influenced by the literature is not a trivial one, but it is not only the literature which affects the researcher's perceptions and potential biases. In this respect it is worth mentioning a remark made by Ray (1993):

> *We are beginning to realize that if we don't believe in something, it doesn't exist - no matter how much data is thrown in front of us.*

Some academics would argue that researchers come to believe in an idea by reading about it in the literature and by reflecting on what they have read.

In addition the issue of the literature review has direct implications for the Grounded Theory concept of theoretical sensitivity. This is also a problematic concept in that this type of sensitivity can predispose a researcher to follow a previously established train of thought. It is difficult to know how theoretical sensitivity could be acquired if the researcher was not fully conversant with many if not most theories related to the topic, or at least the field of study, being researched.

7.5   The use of theoretical sampling

The concept of theoretical sampling is clearly a central issue in Grounded Theory which understands that theory development occurs over a period of time during which ideas are conceived, explored, refined, perhaps rejected and reconceptualised and further explored and developed. This is of course what the dialectic actually is. In the case of Grounded Theory the arguments created need to be fuelled by empirical research. Thus after the first set of interviews have been acquired, and the data coded categories begin to emerge, it will be necessary to carry on looking for further empirical evidence to enrich the understanding of these issues and categories. Memoing is often cited as the vehicle which connects the first pass at the categories to theoretical sampling in that the researcher describes in memos what type of additional data would be helpful in elaborating the categories and developing the theory. At this point the researcher needs to find data sources which will directly help his or her further understanding of the issues and categories that are emerging. Thus theoretical sampling is a form of purposeful sampling whose purpose is to learn more about the emerging issues and categories (Charmaz 2006).

Pragmatism again underpins this thinking. The researcher want to directly access the experiences and thus acquire the data from those who have something important to say on the subject being studied.

It is necessary to remember that this is to be done with the constant comparison mind set and it will be recognised that this theoretical sampling could be seen as having some similarity to data triangulation.

7.6 Summary

Academic research is particularly challenging on several levels and newcomers to it often find it difficult. There are conceptual problems; there are vocabulary problems; there are data access prob-

lems; there are data processing problems and there are interpretation issues. None of these challenges are easy. There is no universal
way of addressing these challenges and every researcher has to find
his or her own way of coping with the research project. In addition
many universities do not give adequate support to novice researchers and thus there is a high non-completion rate for research degrees.

When novice researchers look for a starting point for their research
Grounded Theory can appear to be attractive in that it offers a
framework which is relatively easily understood and followed, at
least at a superficial level. It is for this reason that it has been such a
great success. It is probably the most successful innovation in social
science research in the 20th century. The four characteristics of
Grounded Theory described here would normally be present in any
research project which described itself as employing a Grounded
Theory approach. The extent to which they would be addressed
would of course vary as would the competence with which these issues would be discussed. In a non-Grounded Theory research project
these four issues would at best be very superficially addressed.

Perhaps more important is the fact that the ideas described by Glaser
and Strauss are useful pointers to sound academic research. The
value of Grounded Theory, more than anything else, is the fact that it
opened the black box of qualitative research and allowed the research community to look inside at the concepts that actually drive
academic enquiry. In this sense none of the individual parts of
Grounded Theory matter much. What is much more important is the
mind set Grounded Theory has helped to create. Every academic research project represents the personal voyage of discovery of the

researcher and thus it is inappropriate to over prescribe its actual course. For many it can be good enough to say that the research has been informed by Grounded Theory.

8

An end note on the Nature of Grounded Theory

Academic research is based on the belief that it is possible to understand the world around us. This understanding may come from ideas we have previously acquired or it may result from what we perceive and how we are able to "get behind" our mere perception of phenomena and connect with the principles and theories which explain how the world works. In this sense research is quintessentially an intellectual activity. To achieve understanding, the primary tool at our disposal is our cognitive capacity and this is the driver of all research, but especially theoretical research. However theoretical research alone will not address all the questions for which answers are required for 21st century life. For this we need empirical research and thus primary data. Expressed differently our cognitive capacity is not always robust enough to understand what we are examining and consequently we have the need to acquire appropriate data or evidence to both support and stimulate our thinking. In fact in many if not most cases it is this combination of cognitive capacity and data or evidence which leads to acceptable academic research findings.

The application of the researcher's cognitive capacity and the acquisition of data or evidence are often challenging to even experienced researchers. There are many ways in which a researcher can be misled by either fussy thinking or by apparently appropriate data and it is for these reasons that research methodology is such an important issue as it should signal to us when we are drifting off track.

Grounded Theory is one way of helping researchers face the challenges of academic research. Clearly it is only one of many ways. While focusing on the importance of data and being prescriptive about how it should be treated by the researcher, and at the same time not being too specific about what actually constitutes data, Grounded Theory provides an insight as to what it means to be a dataist. As already mentioned Barney Glaser remarked that "all is data", a comment which has proved challenging to many researchers. In fact this expression is not especially useful other than it reveals a mindset which is required for Grounded Theory. That mindset believes in the paramount importance of the data and that the researcher has to continually revisit the data until a theory or theoretical conjecture emerges and becomes apparent. Of course theories or theoretical conjectures do not emerge on their own, but rather as a result of the cognitive capacity of the researcher which attributes meaning to the data. What makes the dataist approach different is a more intense emphasis on the data and an understanding that the theory or theoretical conjecture will be arrived at through reflection which takes the form of a slow realisation of what the data means.

To only think of Grounded Theory as a method, which was created by Glaser and Strauss and further developed by Corbin, Charmaz and Bryant is to misunderstand the importance of these thinkers' contribution to our knowledge of how academic research should be conceptualised and operationalised. What has been achieved by these individuals is much more important than providing a mere method, in fact as a method Grounded Theory is rather limited. But, by being exposed to the debate surrounding Grounded Theory we are able to reflect on issues which go to the heart of the research process. Knowing what these are allows us to be much better informed researchers with an ability to configure our own research in a way that meets our own particular requirements.

Perhaps what is required is a new term which could be *Grounded Research*, and this could refer to a selection of the thinking of Glaser, Strauss, Corbin, Charmaz and Bryant which is appropriate to answering a particular research question and which is executed in such a way that the research question is answered appropriately.

It is worth remembering that Grounded Theory is greatly admired by some researchers while it is shunned by others and from an academic degree perspective it needs to be engaged in with this diversity of opinion in mind. It is probably wise not to rush to choose a Grounded Theory approach to one's research until other approaches have been evaluated and seen to be unsatisfactory.

In the end research questions are answered by researchers developing a solid chain of evidence which leads through a series of rational steps to a convincing argument and research methods are employed only to assist in this process. Academic researchers forget this at their peril.

Glossary

The principal concepts required for Grounded Theory

Concept	Definition
A code	A code could be a single word or a group of words or even an acronym that will represent the concept being studied. Coding provides a level of abstraction whereby similar ideas are grouped on the basis that they represent a pattern or a theme and are then assembled into high level concepts, categories or constructs.
A paradox	This occurs when two or more conflicting results occur from a logical argument.
A priori	A Latin expression used by researchers meaning, from before. Thus a priori research may not be able to determine how open the chosen informants will be during an interview.
Abduction	A researcher employs abduction when he or she proposes a connection or a pattern between different sets of data or evidence without being fully aware of all that could be learnt about the context and the nature of the data sets. In common parlance one can describe abduction as a type of informed guessing. Abduction is an ugly word coined by the pragmatists and in particular Charles Saunders Pierce.
Academic Masters	This term is sometimes used to describe well established academics that have a reputation for research excellence including the production of new theory.

Concept	Definition
All is data	An expression that points to the fact that researchers have to be always on the lookout for any data which can contribute to a greater understanding of the research question and to a research finding. This expression is credited to Barney Glaser.
Argument	A series of logical statements leading to a convincing conclusion. Arguments are a key aspect of academic research whether it is theoretical or empirical, positivistic or interpretivistic.
Axial coding	A second order form of coding using open codes as its raw material. This involves grouping open codes to constitute more comprehensive constructs.
CAQDAS	Computer Assisted Qualitative Data Analysis Software is a group of products which allow text data to be entered into a computer and to be coded. Once this has been completed the software enables a wide range of analysis to be performed on the data.
Categories	This term is used in Grounded Theory to denote a concept or a variable or even a construct. Categories are established by Grounded Theorists as they code the transcripts of their data collection activities. As the coding proceeds, the categories acquire properties and dimensions and they become the building blocks from which the final theory will be developed.
Cognitive capacity	The ability to understand. People's cognitive capacity is dynamic. The general rule is the more one knows the more one understands.
Confirmatory bias	Refers to a tendency for the researcher to look for data which will confirm his/her preconceptions and to avoid data which may contradict them. This bias will also be seen in terms of how a particular data set is interpreted.

Concept	Definition
Conjectural nature of knowledge	Our knowledge at any point in time should be considered to be our best understanding at that point in time. It is not possible to predict how our understanding will evolve. What can be said is that our cognitive capacity has been evolving for millennia and there is no a priori reason to suppose that it will not continue to do so.
Constructs	A construct is a research concept which is more complex than a variable, for example leadership is a construct. The issue here is that leadership is a complex attribute which will need to be described in terms of a number of variables.
Contingent nature of knowledge	An important philosophical stance in research is the fact that science is unable to prove anything in an absolute way. What we know scientifically is always subject to possible revision. We do however know that we do not know what we do not know. This means that we may discover new laws of science which will turn much of what we now think we know on its head.
Continuous comparison	One of the basic principles on which Grounded Theory has been developed is that data on its own does not reveal that much insight but that data from one situation compared to another situation can be quite revealing.

This is the same principle which is frequently required when using data to comment on performance. For example, if one knows that the ROI is 10%, can it be said whether that return is good or bad? If other similar organisations are earning only 5% then it is probable that the 10% is good but if other similar organisations are earning 15% then it is probable that the 10% is not good. Grounded Theory needs the data to be understood in terms of the context of what is happening around it. |

Concept	Definition
Data	Data are numbers, words, or images or sounds or other sensory stimuli which cause a researcher to take notice and which he or she may or will use in his or her research process. Data needs to be contrasted with information which is sometimes referred to as data that has previously been processed or structured.
Data coding	The process of labelling variables or concepts. At first glance this appears to be an unproblematic issue but in fact it is central to any research project where coding is being performed.

Once the codes are settled then this describes a data analysis trajectory.

It is most important that researchers be prepared to revive their thinking with regard to their chosen codes as they work with a manuscript or transcript. |
Dataism	The term Dataism is sometimes used to describe the type of approach which is used in Grounded Theory research. This term is used to signal the paramount importance which is attributed to data by Grounded Theory researchers.
Democratised research	Prior to Glaser and Strauss theory generation was perceived as being a special activity and one which only a few privileged members of the academic community could aspire to. The privileged members of the academic community were sometimes referred to as the intellectual capitalists. Grounded Theory allowed almost anyone to claim that they had generated a theory, thus democratising research.
Dialectic	The process of a thought being mulled over by more than one mind and some reservations, or additions, or objectives being raised resulting in the thought being refined.

Concept	Definition
Empiricism	An approach to research which requires the acquisition of primary data, often referred to as sensed data as evidence of the existence, nature and characteristics of the phenomenon being studied.
Enlighten-ment	Sometimes referred to as the Age of Enlightenment which occurred in the 17th and 18th centuries and which was characterised by the rejection of the orthodox view that knowledge was handed down by those in authority and by the demand for knowledge to be based on logic and empiricism.
Ethnography	A qualitative research approach which requires the researcher to become familiar with the lived experiences of the research subject/s. This technique is longitudinal in nature and thus requires the availability of a long period of evidence collection and understanding.
Falsification	A term popularised by Karl Popper which asserts that for a statement to be scientific it has to be expressed in such a way that it can be falsified i.e. rejected after close examination.
Field Notes	Personal notes that researchers make for themselves over and above the main collection of data from the informant. These are often quite informal.
First cycle codes	At the end of coding a transcript there may be many dozens or scores of codes. Also codes could overlap or be quite similar. There is often a need for additional codes using higher level concepts for the second coding. The codes used during the first pass through or analysis of the original transcript may be referred to as the first cycle codes.

Concept	Definition
Fit	Fit describes how well a substantive theory corresponds to the situation it describes.
	The concept of fit is problematic in that there are questions to be asked such as how to assess if there is a fit and how much fit should there be before a theory should be regarded as having fit.
Formal theory	A theory which is applicable to a number of different situations and/or organisations and/or individuals. Thus formal theory has a wider scope than substantive theory.
Fracturing the data	In qualitative research data may be viewed though a number of perspectives. One such perspective is to see the data as a holistic entity and the opposite of this is to view the data as a series of different issues.
	When a researcher views the data as being evidence related to a series of issues it is sometimes said that the data is being fractured. In Grounded Theory this is the principal way of working with data. The process of fracturing is represented by the coding of the data where important issues and concepts are identified, labelled (i.e. coded) and then grouped into categories.
	The opposite approach is to understand the data as a whole through the use of some form of hermeneutic analysis.
Grab	Grab refers to how well the theory is accepted by the community in which it exists and for which it was developed.
Grand Theory	A term used to describe theory from former times. Grand Theory would have been postulated by authorities like Smith, Jung, Durkheim, Marx, Weber and other academic Masters.

Concept	Definition
Grounded Theory	*Grounded theory is an inductive, theory discovery methodology that allows the researcher to develop a theoretical account of the general features of a topic/situation while simultaneously grounding the account in empirical observations or data.* (Glaser and Strauss 1967).
Group re-search	A number of researchers working together to facilitate activities such as establishing robust codes. A group will probably lead to richer insights when exploring the intricacies of the data. When a theory is being created the existence of a group implies a greater facility for the employment of the dialectic.
Hermeneutics	A holistic approach to data analysis which is sometimes contrasted with fracturing the data.
Holistic	A term coined in the early part of the 20th century. In research it is used to describe an approach which finds reductionism inappropriate.
Hypothesis	A hypothesis is a claim or an assertion which has been deduced from an established theory the researcher wishes to test. Hypotheses are tested in order to ascertain if they can be rejected. If they are not rejected they are considered to be valid pro tem.
Induction	The process of abstracting from an incident or phenomenon or a series of incidents or phenomena a general principle or theory which will describe the nature of the incident or phenomenon incorporating the issues, variables or constructs present in the incident. Induction can also be described as the intellectual process or a method of inference in which the researcher moves from data to theory.
Integrity of this data	When data is reported in the research its provenance is not often questioned. But there are a number of ways in which the integrity of the data could be questionable, including was the data collected from appropriate knowl-

Concept	Definition
	edge informants? What were the conditions under which the data was collected? Was the data stored securely? Was the transcription conducted competently? Etc.
Intellectual capitalist	Well established academic researchers who are regarded as being competent to deliver quality academic research including the development of new theories. See:- Democratised research .
Interpretation	The process of associating meaning to data in such a way that it will be useful to the researcher.
Interpretivist	An individual who practices Interpretivism, which is often seen as a research tradition competing with positivism.
Joint research project	Research projects which are undertaken on a collaborative basis by different researchers or groups of researchers. Except research conducted for degree purposes much of academic research is undertaken in this way.
Knowledge	There are many definitions of knowledge but the Grounded Theorist will normally choose the Pragmatist definition which relates to the application of knowledge and what can be achieved with it. In the context of Grounded Theory the idea that knowledge solves problems may be traced back to Charles Sanders Peirce and the Pragmatists.
Labelling	The word labelling is normally associated with coding. Coding or labelling concepts or phenomena in order that they may be grouped into categories and from there developed into the building blocks of theory.
Motherhoods	Statements of the obvious or platitudes which are unworthy of being included in the findings of academic research.

Concept	Definition
Open coding	The attribution of codes (numeric or alpha-numeric) in order to fracture or parse data in order to be able to summarise the findings of research. Open coding is the term used to describe codes which are allocated during the first pass through the data.
Originality	Academic research aspiring to be the basis for obtaining a doctorate is required to be original. In this context original-ity is difficult to define but it is characterised by producing a finding which is novel. A question often asked is *How novel?* It is difficult to answer such a question except to say that provided the research is able to bring a new point of view to the subject then this is usually enough. See:- Some-thing of value to the body of knowledge.
Outlier	A data point which is far from all other data points ob-served. In the context of Grounded Theory an outlier could be an informant who has provided significantly different evidence to all the other informants who have taken part in the study.
Perceptual maps	A visual representation of data in such a way that the data relationship with other variables may be clearly perceived by the researcher.
Persistent interaction with the data	Grounded Theorists are expected to continually interact with the data in order to obtain a full understanding of the subject being researched. No timeframe can be suggested for this period of persis-tence.
Positivist	A research mindset which has many characteristics includ-ing but not limited to requiring the researcher to be a real-ist; a belief that the researcher can be unbiased. Positivists normally work within a hypothectico-deductive framework where there is a general belief of cause and effect. Much of the work of a positivist is performed using quantitative

Concept	Definition
	data and this is incorrectly sometimes thought to be the defining characteristic of positivism. The objective of this approach to research is often said to be prediction and control.
Pragmatism	A philosophical position taken by researchers and named by Charles Sanders Peirce that says what works is in some sense right. Also what is useful is "true" or right. This philosophy was further developed by William James and John Dewey. According to the pragmatists the practices of inquiry is a social activity.
Primary data	Data acquired by the researcher as part of the research process. Primary data is often contrasted with secondary data which is data that has already been collected by others and exists in a published form.
Problem-solving ethos	Social science research is normally driven by a need to solve a problem and the findings are judged to the extent that the problem is solved or ameliorated by the application of the knowledge produced in the findings. In such cases it is said that a problem-solving ethos is at work.
Prolonged emersion	In order to develop a theory academic researchers need to the thoroughly acquainted with the situation in which they are arguing the theory will apply. This takes a long time and requires the researcher to ensure that he or she really understands the environment, which may take months and is referred to as having a prolonged emersion in the environment being studied. It is worth noting that when ethnographic studies are involved then the researcher could find the period of emersion to be even longer.

Concept	Definition
Proposing a theory	The objective of Grounded Theory is to develop a new or novel theory. Thus the final output of a Grounded Theory research project will be the proposing of a new theory. This is sometimes referred to as a theoretical conjecture and it sometimes remains a conjecture until further work has been done to strengthen its validity.
Protestant persuasion	The term Protestant persuasion alludes to the Protestant Ethic which was an expression coined by Max Weber who argued that spiritual salvation and worldly prosperity could be achieved by the work and thrift ethic. For this reason it was argued that on average those living in Protestant ori-entated societies enjoyed a higher standard of living that the average person in non-Protestant societies.
Psychological dimension to induction	Induction is a powerful research inference tool. However when considered carefully there is no logical basis for the use of induction. It simply does not follow that if some-thing happens every day that it should be believed that it will happen tomorrow. Nonetheless we do so believe, and one explanation of this is that at a profound psychological level human beings need to believe that tomorrow will be like today.
Pure research	Research which is conducted for its own sake and which does not promise to answer a current problem. There are few if any opportunities for pure research in the social sciences.
Purposeful sampling	One in which the elements have been deliberately se-lected for their properties.
Puzzle solvers	An expression coined by Thomas Kuhn to describe research undertaken on a routine basis. Kuhn points out that there are few breakthroughs which change fundamental thinking on a routine basis.

Concept	Definition
Qualitative research	Research which primarily uses words and images as the primary data source. Some numeric data such as tables and graphs may be used without changing its status as qualitative research.
Random sampling	The selection of data points of informants from a larger population in such a way that each selection could have chosen any data point or informant in the population. Every data point and informant has an equal probability of selection.
Reflection	The process of introspection on various aspects of the research, such as what do the results of the research actually mean? Reflection is also used to explore the motives of the researcher and the choices which have been made during the research process.
Research Memo	A note which the researcher writes to himself or herself. In Grounded Theory this is a formal issue and Research Memos are seen as critical to good research practice.
Research question	This is the raison d'etre of the research. The researcher tries to find an answer to the research question. Without a research question academic research has no direction.
Rigour	Academic rigour refers to research being conducted in such a way that the academic community has confidence in the results. This will normally mean that whatever method has been chosen the rules and procedures associated with that method have been followed carefully.
Rigorous method	An approach to academic research which has been carefully thought out and which inspires the community to have confidence in the research findings.
Schism	A split or a division or an argument usually among the clergy or scientists about matters of principle. The term is usually reserved for serious differences.

Concept	Definition
Scholarship	A characteristic involving being well read and also being able to use such knowledge to create convincing arguments in the scholars' field of interest. Scholarship requires skill at critique as well as at assembling new concepts and skilfully presenting them to peers for review and comment.
Secondary data	See Primary data.
Second cycle coding	Text is coded for the purposes of grouping ideas. However sometimes these codes need to be further grouped. When this is done it is referred to as second cycle coding. Sometimes third cycle coding may also be required.
Something of value to the body of knowledge	One of the primary objectives of academic research especially at the doctoral level is that the research findings should add something of value to the body of knowledge. This is sometimes described as there needing to be an original finding. However the word original is difficult and some academics suggest that it is better to point out that the findings have to say something new and this has to be of some significance to the academic community. It is not easy to define this but academic research is sometimes described as being a voyage of discovery. Marcel Proust made a comment which may help academics understand what is required. He said "The real voyage of discovery consists not in seeking new landscapes, but in having new eyes".
Substantive Theory	A theory which is applicable to the institution or the individuals who were participants in the research process. This type of theory would not be necessarily generalisable to a larger population.
Tabula rasa	This is a Latin expression used by researchers. Its literal translation means, a clean slate.

Concept	Definition
Theoretical conjecture	A theoretical statement or even a theory which has yet to be tested. Researchers will normally create a theoretical conjecture which the research will test. Theoretical conjectures may be stated by a researcher as part of a thought experiment by which the researcher looks for feasible explanations of phenomena.
Theoretical Sampling	A form of purposeful sampling whereby researchers find data from suitable cases/participants/ organisations to strengthen the researchers' understanding of the categories and constructs emerging from the research. Theoretical sampling may be thought of as a sort of triangulation.
Theoretical Saturation	When additional informants do not provide any new data or insights a point of theoretical saturation has been reached and there is no value in attempting to access additional informants. This can also be referred to as data saturation. The number of informants needed for saturation varies substantially from one research project to another.
Theoretical Sensitivity	A term used to describe a researcher's experience and knowledge which he or she will draw upon in moving from data collection and analysis to theory generation. Theoretical Sensitivity requires a substantial level of maturity and an ability to be reflective. It requires a sense of what is likely to be credible. It requires an ability to be self critical.
Theory	Systematically organised knowledge applicable in a relatively wide variety of circumstances, using a system of assumptions, accepted principles and rules of procedure devised to analyse, predict or otherwise explain the nature or behaviour of a specified set of phenomena. But it is also often simply the best explanation which is available at that time.

Concept	Definition
Theory creation	There are many different ways of creating theory but it is useful to clarify again that there are different ways of thinking about what theory actually is. Einstein (1950) definition is helpful. He said that, *"Science is the attempt to make the chaotic diversity of our sense-experience correspond to a logically uniform system of thought. In this system single experiences must be correlated with the theoretical structure in such a way that the resulting coordination is unique and convincing"*.
Theory testing	An important function of academic research is to test theories. This is done through a structured approach which requires hypotheses to be derived or deduced from the theory. The researcher then seeks data which will facilitate his or her being able to reject or discard the hypotheses or claims. It is important to note that theory testing cannot prove a hypothesis or a claim. If the hypothesis is not rejected it is accepted pro tem. If the hypothesis is rejected the research then amends the theory to reflect the new knowledge so acquired.
Triangulation	The word triangulation has been borrowed from land surveyors. It refers to the use of multiple lenses through which to consider most aspects of the research process. There are several ways in which triangulation may be used such as data triangulation, informant triangulation, method triangulation to mention only three. Triangulation uses multiple approaches or tools or data in order to obtain a greater understanding of the phenomenon being studied. Sometimes triangulation is seen as a method of cross-checking the credibility or validity of that is being discovered, but more correctly it offer a richer and more complete understanding of the phenomenon.
Truth	A difficult concept to define, truth is that which is believed to have happened or to exist. It may also be about the fu-

Concept	Definition
	ture. When something is said to be true or to represent the truth then there is no doubt about it. There are many issues about which it is difficult to challenge a statement said to be true such as, the name of the America President in 2013 is Barak Obama or water boils at 100 degrees centigrade at one atmospheric pressure. However if the water is impure or there are unusual conditions such as humidity, water will boil at a different temperature.
Uncertainty	Most aspects of academic research have a degree of uncertainty associated with it. There can be uncertainty about the integrity of the data, the appropriateness of the analytical techniques and about the inferences used in moving from the data to the findings. A belief in certainty is a sign of not understanding the academic research ethos.
Variables	The characteristics being studied are regarded as variables.

Reference List

Ackoff R, (1989), From data to wisdom, Journal of Applied Systems Analysis 16, p 3–9.

Feyerabend P, (1990), Against Method, p18, 3rd Ed, Verso, LondonFeynman R, (1995), Six Easy Pieces, p2, Penguin Books, London Raimond P, (1993), Management Projects, p93, London, Chapman & Hall.

Alvesson M and Kaj Sköldberg, 2008, Reflexive Methodology: New Vistas for Qualitative Research, Sage, Thousand Oaks

Anonymous, (2013), Pragmatism, http://www.youtube.com/watch?v=tRxTb6JPVHY

Ashall F, (1996) *Remarkable Discoveries*, Cambridge: Cambridge University Press.

Babbie E & J Mouton, (2001), The practice of social research. Cape Town: Oxford University Press.

Birks M and J Mills, (2011) Grounded Theory - A practical guide, Sage, London

Born M, (1988), cited by Gerald Holton's *Thematic Origins of Scientific Thought*

Broad W and N Wade, (1985), Betrayers of the Truth: Fraud and Deceit in Science, Oxford University Press, Oxford

Bryant A, (2009), Grounded Theory and Pragmatism: The Curious Case of Anselm Strauss, FQS, Vol. 10, 3, Art 2, Septemberhttp://www.qualitative-research.net/index.php/fqs/article/view/1358/2850, accessed February 20, 2013

Burke J, (1985) *The Day the Universe Changed*, Boston: Little, Brown & Company.

Carroll L, (1982) *Alice's Adventures in Wonderland*, first published in 1872, London: Chancellor Press.

Charmaz K, (2006), Constructing Grounded Theory, Sage, London

Checkland P, (1986), *Systems Thinking, Systems Practice*, Chichester: John Wiley and Son.

Clark V and J Creswell, (2008), The Mixed Method Reader, Sage, Thousand Oaks, CA

Collins H, (1975), 'The Seven Sexes: A Study in the Sociology of a Phenomenon, or the Replication of Experiments in Physics', *Sociology*, Vol. 9, No. 2, pp. 205–24.

Collins H, (1994), A broadcast video on science matters entitled *Does Science Matter?*, UK: Open University/BBC.

Conan Doyle A, (1992), The Memoirs of Sherlock Holms, Wordworth Classics, Ware

Cooksey R, (2012), Measurement Issue, http://www.youtube.com/watch?v=hTwqYeq1DPo viewed February 22, 2013

Corbin J and A Strauss, Basics of Qualitative Research 3e, (2008), Sage, Thousand Oaks.

Creswell J. W, Plano Clark, V. L., Gutmann, M.L., & Hanson, W.E., (2003), Advanced mixed methods research design. In A Tashakkori and C Teddlie (Eds.), Handbook of mixed methods in social and behavioural research, 209–240. Thousand Oaks, CA: Sage.

Darwin C, (1978), cite in, The Voyage of the Charles Darwin, BBC Productions.

Darwin C, (1986), *The Origin of the Species*, first published in 1859, London: Pelican Classics.

Denzin N and Y Lincoln, (Eds), (2003), Handbook of Qualitative Research, , Sage, Thousand Oaks, CA.

Dewey J cited by Thayer H., (1970), Pragmatism, the Classic Writings, Hackett Publishing, Indianapolis.

DiMaggio P, (1995), 'Comments on "What Theory is Not"', *Administrative Science Quarterly*, Vol. 40, pp. 391–7.

Einstein, A. (1950) 'The fundamentals of theoretical physics', in Out of my Later Years, Philosophical Library, New York.

Eliot, T.S., 'Little Gidding', *Four Quartets*, cited in E. Knowles (1999) *Oxford Dictionary of Quotations*, Oxford: Oxford University Press.

Feyerabend P, (1993), Against Method, p.39, 3rd Ed., Verso, London.

Feynman R, (1995), *Six Easy Pieces*, London: Penguin Books.

Ghoshal S,(2005), Bad Management Theories Are Destroying Good Management Practice, Academy of Management Learning and Education, 005, Vol. 4, No.1, 75-81.

Gibbons M Camille Limoges, Helga Nowotny, Simon Schwartzman, Peter Scott, & Martin Trow, (1994), The new production of knowledge: the dynamics of science and research in contemporary societies. London: Sage.

Giddens A, The consequences of Modernity, Polity Press, p39, Cambridge, 1990.

Glaser B & A Strauss, (1967), The Discovery of Grounded Theory: Strategies for Qualitative Research, Chicago, Aldine Publishing Company.

Gould SJ, (1992),The Mismeasure of Man, p27, Penguin Books, London

Gribbin J, (2002), *Science: A History*, London: Penguin Books.

Gummerson E, (2000), Qualitative Methods in Management Research, Sage, Thousand Oaks

Habermas J., (1993), Postmetaphysical Thinking - Philosophical Essays, p.6, translated by Hohengarten, The MIT Press, Cambridge Massachusetts.

Honderich T, (1995), *The Oxford Companion to Philosophy*, Oxford: Oxford University Press.

Horgan J, (1996), The End of Science, p 202, Abacus, London, citing N Oreskes, K Belitz and K Sharader-Frechette from Verification, Validation, and Confirmation of Numerical Models in the Earth Sciences, Science, Feb 1994, p 641-646.

Ishiwara J in Einstein The e in Koen has a bar over the top of it Koen-Roku, Kyoto Lecture reported by (Tokyo-Tosho, Tokyo, 1977) cited by A. Pais in Subtle is the Lord: The Science and the Life of Albert Einstein (OUP, 1982) p. 179.

Kelvin Lord, formerly William Thomson, Found at http://www-history.mcs.st-andrews.ac.uk/Quotations/Thomson.html, accessed 14 August 2013

Kennedy M, (1979), Evaluation Quarterly, Vol 3, No 4, November, p 661-678

Keynes J, (1936), The General Theory of Employment, Interest and Money, London: Macmillan.

Locke, J. (first published in 1690), 'An Essay Concerning HumanUnderstanding', in J. Taylor (ed.) (1974) *The Empiricists*, New York: Anchor Books.

Medewar P, (1986), *The Limits of Science*, Oxford: Oxford University Press.

Menand L., (2002), The Metaphysical Club, Flamingo, London.

Nietzsche F., The Will to Power, Wille zur Macht 481. Cf. Stanford Encyclopedia Philosophy: http://plato.stanford.edu/entries/nietzsche/ also Vintage Books, New York, 1967.

Paulos J, (1998), Once Upon A Number, p14, The Penguin Press, London

Plato (Originally published circa 380 BCE), in G.M.A. Grube and C.D.C. Reeve (trs) (1992) *The Republic*, Indianapolis: The Hackett Publishing Company.

Ray M, (1993), Introduction: What is the new paradigm in business? In The New Paradigm in Business, p5, G P Putnam's Sons, New York,.

Remenyi D and G Onofrei and J English (2011), An Introduction to Statistics Using Excel, Academic Conferences, Reading

Remenyi D, (2012), Field Methods – Interviews, Focus Groups and Questionnaires, Academic Publishing International, Reading

Remenyi, D., Williams, B., Money, A. and Swartz, E. (1998) *Doing Business Research*, London: Sage.

Rosenthal R and R Rosnow, (1991), The Essentials of Behavioural Research: Methods and Data Analysis, McGraw Hill International series, Second Edition, New York.

Russell B, (1960), Our Knowledge of the external world, Mentor Books by New American Library of World Literature, NY

Sacks, O. (1991) *A Leg to Stand On*, London: Picador-Pan Books.

Scott Fitzgerald F K, 1920, in The Author's Apology, a letter to the Booksellers' Convention,(published in The Letters of F. Scott Fitzgerald, ed. by Andrew Turnbull, 1963), referring to his novel This Side of Paradise.

Shubik, M. (1988) *A Game-Theoretical Approach to Political Economy*, Cambridge (MA): MIT Press.

Strauss and Corbin, (1998), Basics of Qualitative Research, Sage, Thousand Oaks.

Sutton, R.I. and Staw, B.M. (1995) 'What Theory is Not', *Administrative Science Quarterly*, Vol. 40, pp. 371–84.

Urquhart C, (2013), Grounded Theory - A practical guide, Sage, London

Wheatley M, (1992), Leadership and the New Science, p.8, Berrett-Koeler Publishers, San Francisco.

Whetten D, (1989), 'What Constitutes a Theoretical Contribution?', *Academy of Management Review*, Vol. 14, No. 4, pp. 490–5.

Whewell W, (1964), cited by Sir Peter Medawar, BBC Talk, first published1963: Is the scientific paper a fraud?. Listener 70, 1963 Reprinted in: Medawar P. (1991) The Threat and the Glory Oxford: Oxford University Press, pp. 228-233 also reprinted in: Peter B. Medawar. The Strange Case of the Spotted Mice, OUP. 1996

Winston, B. and Fields, D. (2003) 'Developing Dissertation Skills in an Internet Based Distance Education Curriculum: A Case Study', *American Journal of Distance Education*, Vol. 17, No. 3, pp. 161–72.

Wittgenstein L.,(1969), Wittgenstein On Certainty, sct. 378 (ed. by Anscombe and von Wright.

Wolpert L,(1993), *The Unnatural Nature of Science*, London: Faber and Faber.

Index

www.ingramcontent.com/pod-product-compliance
Lightning Source LLC
Chambersburg PA
CBHW070910270326
41927CB00011B/2512